T0251886

Restructuring Global Regional Agricultures

Transformations in Australasian agri-food economies and spaces

Edited by
DAVID BURCH
Griffith University, Australia
JASPER GOSS
Griffith University, Australia
GEOFFREY LAWRENCE
Central Queensland University, Australia

Routledge
Taylor & Francis Group

LONDON AND NEW YORK

First published 1999 by Ashgate Publishing

Reissued 2018 by Routledge
2 Park Square, Milton Park, Abingdon, Oxon, OX14 4RN
711 Third Avenue, New York, NY 10017, USA

Routledge is an imprint of the Taylor & Francis Group, an informa business

Publisher's Note
The publisher has gone to great lengths to ensure the quality of this reprint but points out that some imperfections in the original copies may be apparent.

Disclaimer
The publisher has made every effort to trace copyright holders and welcomes correspondence from those they have been unable to contact.

A Library of Congress record exists under LC control number: 98074503

ISBN 13: 978-1-138-32888-4 (hbk)
ISBN 13: 978-1-138-32900-3 (pbk)
ISBN 13: 978-0-429-44835-5 (ebk)

Contents

Figures and tables

Figures

Tables

Notes on contributors

Christina I. Baldwin is a graduate of the University of Waikato, Hamilton. Formerly a school teacher, she has a masters degree in the sociology of education and feminist history. She is currently a doctoral candidate researching the New Zealand dairy industry from a feminist perspective. She resides and works on a dairy farm in Putaruru, South Waikato. She is a supplier of milk and shareholder in the New Zealand dairy industry, and is actively involved in dairy industry politics.

Greg Blunden is a research fellow in the Department of Geography at the University of Auckland. He is currently involved in co-directing a Public Good Science Fund project on the sustainability of land-based production and rural communities in Northland. He has published recently in *Environment and Planning A* and *New Zealand Geographer*.

Ben Bradshaw is a doctoral candidate in the Department of Geography at the University of Guelph, Ontario, Canada. He researches the environmental implications of the removal of subsidies in commercial agriculture, focussing on new trade regimes such as the World Trade Organisation and the North American Free Trade Agreement. Further research includes examining new stresses and influences on resource management decisions at the farm level. His publications have appeared in *Agriculture, Ecosystems and Environment* and *New Zealand Geographer*.

David Burch is Associate Professor in the School of Science and Director of the Science Policy Research Centre, Griffith University, Brisbane. He is the convenor of the Agri-food Research Network, and co-edited *Globalization and Agri-food Restructuring* (Avebury, 1996), and *Australasian Food and Farming in a Globalised Economy* (Monash University, 1998). His most recent book is the co-authored textbook *Science, Technology and Society: An Introduction* (Cambridge, 1998). His research interests focus on agri-food restructuring in Southeast Asia and resource use in Australian agriculture.

Hugh Campbell is a lecturer in Social Anthropology at Otago University, Dunedin. He has been a long-term member of the Agri-food Research Network and has conducted research into a range of issues surrounding the restructuring of land-based production in New Zealand.

Ian Carter is Professor of Sociology at Auckland University, where he teaches a course entitled *Consuming Passions: the Sociology of Food.* The second edition of his book *Farm Life in North East Scotland, 1840-1914: the Poor Man's Country* was published by John Donald in 1997.

Richard Clark originally trained in veterinary science and is currently Principal Extension Officer (Development) with the Rural Extension Centre of the University of Queensland. He is presently based in Rockhampton where his work involves new action-learning approaches to change among producers. After successfully developing his 'local best practices' model in the beef sector, he has recently applied the approach to cotton and other primary industries.

Lyn Collie holds an honours degree in social anthropology from Otago University. She is interested in the construction of femininity and masculinity through social interaction and media texts, the sociolinguistics of male and female discourse, and the symbolic significance and political economy of food.

Brad Coombes is a lecturer in Environmental Studies and Resource Management in the Department of Geography, University of Auckland. He has research interests in sustainable agriculture, ecotourism, critical perspectives on sustainable development, environmental regulation and Mäori land use policy. He recently completed his doctorate at Otago University on the implications of globalisation for sustainable tourism and Mäori development, and has recently published in *Rural Sociology* and *Sociologia Ruralis.*

Bruce Macdonald Curtis is a lecturer in the Department of Sociology at the University of Auckland. His research interests include the restructuring of agricultural sectors, in particular the meat industry; the development of gaming as a sector of economic and social significance, marketing boards and cooperatives as organisational forms in New Zealand and theories of governance.

Michael Finemore holds a BA(Honours) degree and is enrolled in a Master of Arts degree at Central Queensland University. His research interests lie mainly in the field of rural sociology, and his current work is concerned with sugar industry restructuring, and the social relations and practices of sugarcane harvesting in the Bundaberg district of Queensland.

Gerard Fitzgerald is an anthropologist and environmental sociologist, and the principal of Fitzgerald Applied Sociology, a consulting company located in Christchurch. His work has focussed on the social aspects of resource management, particularly social assessment and participatory planning.

Jasper Goss is a PhD student at the Science Policy Research Centre, Griffith University, Brisbane, currently completing research on agri-food restructuring in Thailand. His interests include the political economy of development, histories of decolonisation and socialism during the Cold War and social movements in Southeast Asia. His publications have appeared in *Third World Quarterly*, *Culture & Agriculture* and *Third World Resurgence*.

David Grasby is a PhD candidate with the Rural Social and Economic Research Centre at Central Queensland University, and is affiliated with the Cooperative Research Centre for Sustainable Sugar Production in Townsville. His research interests include issues surrounding sustainable development and the political economy of the Australian sugar industry.

Vaughan Higgins is a PhD student attached to the Rural Social and Economic Research Centre at Central Queensland University. He is currently investigating the changing role of the Australian state in rural policy, focussing specifically on issues of farm adjustment. His other research interests include rural political power and rural development policy.

Geoffrey Lawrence is Foundation Professor of Sociology and Executive Director of the Institute for Sustainable Regional Development at Central Queensland University. He has completed national studies into aspects of structural change in farming, scientists' and consumers' attitudes to genetic engineering, and rural social disadvantage. His most recent co-edited book is *Altered Genes - Reconstructuring Nature* (Allen and Unwin, 1998). He has held visiting professorships in the US and Britain and is the Oceanic representative of the International Rural Sociological Association.

xiv *Restructuring Agricultures*

Richard Le Heron is Professor of Geography at the University of Auckland. He has a major interest in the changing geographies of global commodity systems. Building on his book *Globalized Agriculture, Political Choice* (Pergamon, 1993), he has explored theoretical and empirical issues relating to globalisation and agri-food restructuring. Two recent co-authored books *The Asian Pacific Rim and Globalization* (Avebury, 1995) and *Changing Places, New Zealand and the Nineties* (Longman Paul, 1996) relate to these themes. He is currently involved in three research programmes funded by the New Zealand Foundation for Research Science and Technology, and concerned with Asia-Pacific economic cooperation, sustainable land-based production and the dairy and sheepmeat sectors in New Zealand.

Ruth Liepins is a rural social geographer and lecturer in Geography at the University of Otago. Her work focuses on community-based research and post-structural approaches to gender and agriculture. She has published in *Rural Sociology* and contributed to the edited collection *Democratisation and Women's Grassroots Movements* (Indiana University Press, forthcoming).

Stewart Lockie is a lecturer in rural and environmental sociology at Central Queensland University and is coordinator of the Rural Social and Economic Research Centre's programme on agri-food restructuring and the environment. His research interests include the environmental impacts of agriculture, the relationships between food consumption, self-identity and the construction of space, and social impact assessment. He is co-editor of *Critical Landcare* (Centre for Rural Social Reseach, 1997).

Jason Mabbett is a PhD student in the Department of Sociology at the University of Auckland. He is currently investigating the New Zealand wine industry. His research interests are in the sociology of agriculture, contract farming, farm families, the role of transnational corporations and the environmental and social implications of agribusiness.

Jim McAllister is a lecturer in sociology at Central Queensland University. His main research interest is in wage labour in rural Australia, focussing in particular on wage workers in primary industries in the Fitzroy River basin, farm workers in a changing agriculture, and agricultural employees in the developing countries of the Pacific Rim.

Megan McKenna has been working in the Department of Geography, Massey University, since the completion of her PhD at the University of Ottawa in 1996. Her current research focus relates to socio-economic change in New Zealand's primary industries. Related areas of interest include gender issues and geographies of development and underdevelopment. She is currently a Research Fellow under the New Zealand Foundation for Research, Science and Technology programme and has published in a wide range of academic and professional journals.

Philip McMichael is Professor of Rural and Development Sociology, and director of the International Political Economy Programme, at Cornell University, Ithaca. He is author of *Settlers and the Agrarian Question* (Cambridge, 1984) and *Development and Social Change* (Pine Forge Press, 1996), and editor of *The Global Restructuring of Agro-food Systems* (Cornell University Press, 1994) and *Food and Agrarian Orders in the World Economy* (Praeger, 1995). His research is concerned with the politics of change in the world capitalist economy, and regionalisation of the Pacific Rim food system. Currently he is President of the Sociology of Agriculture and Food Research Committee (RC-40) of the International Sociological Association.

Melissa Meyers is the research officer for the Land and Water Resources Research and Development Corporation-funded 'Cotton Best Practices' project, at the Rural Social and Economic Research Centre, Central Queensland University. She has an honours degree in Sociology and is currently conducting PhD research which explores the construction of 'youth' identities in regional Australia.

Andy Monk has a BSc from Melbourne University and has recently submitted a PhD to Wollongong University in Australia, entitled *Sustaining Organic Agriculture in Australia*. He works as an adviser on quality assurance and organic production in the food industry and is an international inspector for the organic industry.

Carolyn Morris is a PhD student at the Department of Anthropology, University of Auckland, working on issues of identity and history among high country sheep farming women of the MacKenzie Basin, New Zealand. Her research interests include gender, community, kinship and food. She has also researched the social impacts of dairy factory amalgamations, technology transfer and land use change.

Bill Pritchard is a lecturer in the Division of Geography at the University of Sydney. He is currently researching the spatial strategies of transnational agri-food corporations and the contemporary restructuring of the Australian wine industry.

Khyla Russell is a post-graduate student in the Department of Social Anthropology at Otago University. She has long term links with the New Zealand fishing industry and Otäkou Fisheries, in particular through her *whänau*, some of whom are still involved in the areas of management, fishing and processing.

Michael Roche is Associate Professor of Geography in the School of Global Studies, Massey University. An historical geographer, he has published in the areas of forestry and soil conservation, as well as on aspects of present day agricultural deregulation and restructuring in New Zealand. His current research concentrates on the frozen meat industry between the wars, and the pip fruit industry. His work has appeared in *Geoforum, Progress in Rural Policy and Planning* and *Economic Geography*.

Ruth Schick is an educational researcher and lecturer in Women's Studies at the University of Otago. Her research focuses on issues of literacy and relationships between educational and occupational stratification.

Roger Wilkinson is a social scientist employed by Landcare Research, Lincoln, the Crown Research Institute in New Zealand which specialises in sustainable management of land and fresh-water ecosystems. His research activities include measurement of attitudes, beliefs and behaviours, community participation in natural resource management, and technology and information transfer.

Foreword

The Agri-food Research Network was established in 1993 as a loose affiliation of Australian and New Zealand social scientists interested in food, agriculture and social change. Since its inception, the Network has carried out an investigative, communication and publication programme, focussing on the analysis of the experiences of the Australasian region during a period of worldwide economic restructuring. Members have attempted not only to provide explanations for changes in farming, commodity processing and distribution, and rural communities, but also to influence agri-food policy through, *inter alia*, submissions to government committees, discussions with food union representatives, interaction with rural producers, discussions with agribusiness representatives, keynote addresses to academic audiences, and the publication of critical works such as this volume.

While originally formed within the 'new political economy of agriculture' school, the approach to research has broadened as new members have joined the Network. Alternative epistemologies and theoretical approaches - particularly in the forms of discourse analysis and the various methods associated with post-structuralism - have added new analytical tools and conceptual frameworks to investigations undertaken by Network members. Such additions are important because they make it possible to better understand the micro-level processes occurring at the farm and community levels. Here, the re-entry of history, geography, anthropology, community studies and political science have complemented the broader structural approaches adopted earlier. They have demonstrated the desirability of understanding change as a complex interaction of cultural, political, spatial and historical dimensions while reinforcing the importance of interdisciplinary approaches. The primary interest of Network members is in problematising the structures and social relations of Australian and New Zealand agri-food systems, a focus which has allowed researchers from different disciplinary and theoretical traditions to interact in a collaborative, collegial and productive atmosphere.

An examination of the regional dimensions of global economic change - an underlying 'theme' of the Network - is of great importance for nations such as Australia and New Zealand. It was apparent from the early research conducted by the Network that, largely as a consequence of the unique position occupied by both countries in the global economy, the experience of change in Australia and New Zealand could provide valuable insights into the processes and local impacts of agri-food restructuring. Both countries are major agri-food and raw materials producers located close to the once-booming markets of East Asia, but placed on the periphery of the dominant producer-countries in North America and Europe. Both Australia and New Zealand have also experienced an unparalleled degree of deregulation and liberalisation of their agricultural sectors, through the withdrawal of state supports and trade protections in the 1980s and 1990s. Processes of deregulation, liberalisation and internationalisation have produced spatially uneven effects, including major transformations of the agri-food sectors of Australia and New Zealand, as global corporations increased their influence over on-farm production, and off-farm marketing, of foods and fibres. Locally-based corporations have also developed conscious strategies to take advantage of these new circumstances, which have included on the one hand, the development of linkages to markets in East Asia and, on the other, a move into organic foods in response to demands for food produced in environmentally-sound ways.

This volume builds on previous Network publications (*Globalization and Agri-food Restructuring*, Avebury, 1996, and *Australasian Food and Farming in a Globalised Economy*, Monash University, 1998), and focusses on issues of continuing relevance to members, as well as issues which have arisen recently. The changes to the region in both environmental and economic terms, for example, are reflected by the attention paid to subjects such as sustainability, organic production, environmental impacts and the role of export markets (see chapters 5, 11, 13, 19). An increasing desire to explore theoretical issues related to globalisation, gender, consumption, technology transfer and the classical debates of agrarian political economy is also demonstrated (see chapters 1, 2, 3, 4, 6, 16, 17, 18). Finally, the local dimensions of change are explored through the critical analysis of government policies (see chapters 7, 8, 9) and local responses to social and biological transformations (see chapters 10, 12, 14, 15).

Further to this process of analysis, the editors have grouped the chapters in this collection into four parts. Each part reflects the overarching theme of the papers, beyond the subject descriptions above. The chapters in Part I attempt to conceptualise a series of global trends and their ramifications as reflected in transformations at local and regional levels. The chapters of Part II seek to interrogate Australian and New Zealand experiences of agri-food restructuring in historic, ideological and discursive terms. Part III analyses local policy and politics, describing the responses to change and the broadening of concerns for rural producers as rural politics becomes increasingly diffused and decentred. Finally, the chapters in Part IV study four key concepts underpinning agri-food research and their continued relevance, and investigate the possibilities for their application in new areas. The analysis of all these issues attests to the importance of the work done by members of the Network, and of the evolution of a research agenda which has significance not only for other researchers, but also for the lived experience of rural communities and economies throughout the region.

The chapters in this volume were initially read as papers at the fifth annual conference of the Network, which was held at Akaroa, New Zealand in December 1997. Our thanks go to the organisers - Hugh Campbell (Otago University), John Fairweather (Lincoln University) and Alison Loveridge (Canterbury University) - whose considerable efforts and dedication resulted in the largest Network meeting to date, and one which was extremely successful, effective and greatly enjoyed. Thanks are also due to Philip McMichael, who made the long journey from the United States to present his invigorating and stimulating keynote paper.

The editors would like to thank the members of the Network who contributed chapters to this volume and who, as always, made the task of editing a very rewarding experience. We are grateful to Eugene Ahn and Eric Metz for their patience in explaining and demonstrating the details of desk-top publishing. We would also like to acknowledge the financial support made available by Canterbury, Lincoln, and Otago Universities, New Zealand; the Science Policy Research Centre, the School of Science, and the Australian School of Environmental Studies, Griffith University, Brisbane; and the Rural Social and Economic Research Centre and the Institute for Sustainable Regional Development, at Central Queensland University, Rockhampton. We would also like to express our thanks to the editorial staff of Ashgate Publishing for their important contribution.

For readers wanting to know more about the Network and its activities, details are available on the World Wide Web, at <http://www.sct.gu.edu.au/sci_page/research/agri>.

David Burch
Jasper Goss
Geoffrey Lawrence

September 1998

PART I
THE CONTEMPORARY
DYNAMICS OF GLOBAL
AGRI-FOOD RELATIONS

1 Virtual capitalism and agri-food restructuring

PHILIP McMICHAEL

Introduction

Developments in the late-twentieth century seem to have reversed our ontological bearings: no longer does money make the world go around, rather it is the world that makes money go around - more than US$2 trillion circulates globally every day.[1] At present the total annual value of global financial transactions is twice the total value of world production, and the real economy in foreign exchange transactions is down to 2.5 percent, with the remaining 97.5 percent being speculative (Korten, 1995; Lietaer, 1997: 7).

We have seen competitive capitalism, monopoly capitalism, late capitalism, and disorganized capitalism, come and go - and it appears that we now have 'virtual' capitalism. Since capitalism is a political form, 'virtual' capitalism has a distinct form of political management, which will be the focus of this essay, and which I shall illustrate through the derivative processes of agri-food restructuring. My goal is to specify the political character of contemporary capitalism, employing a comparative-historical perspective. I contrast the political character of nineteenth and twentieth century capitalism, during the periods of British and US hegemony respectively, as benchmarks from which to specify late-twentieth century 'virtual' capitalism.

Specifying capitalism involves a method of periodising capitalism by its institutional history. This means that rather than understanding capitalism as a series of stages of accumulation, it is necessary to analyse its political configurations and conjuncture. Historical conjunctures emerge as politicised forms of accumulation, anchored in discursively governed configurations of inter-state relations. This definition applies as much to the concept of the Bretton Woods monetary regime as it does to the post-World War Two food regime, which is a sub-set of the monetary regime (and refers to the political management of the world food economy). Such concepts identify or embody sets of definitive historical relationships that serve as points of departure in interpreting (as opposed to predicting) social processes, including agri-food restructuring.

The goal of this essay is to specify the definitive historical relationships currently emerging in the era of 'virtual' capitalism, or 'globalisation.' While I do not believe we can identify a third food regime as such, current agri-food restructuring is governed profoundly by the politics of globalisation. I shall nevertheless begin with the concept of the 'food regime,' a derivative concept concretising the historical relations of capitalism, in order to establish world-historical perspective.

The food regime

The concept of the food regime (Friedmann, 1987) was a child of its time. It manoeuvred deftly between Immanuel Wallerstein's formulaic concept of the world system and Michel Aglietta's concept of regulation, analysing the rise and decline of national agricultures as part of the geo-political history of capitalism. It was not simply about food, but about the politics of food. An initial definition of the 'food regime' was that it linked 'international relations of food production and consumption to forms of accumulation broadly distinguishing periods of capitalist transformation since 1870' (Friedmann and McMichael, 1989: 95). In other words, capitalism was periodised in geo-political terms and its periodisation coincided with two different moments in the life of the nation-state. This analysis featured an interpretation of the historical conditions under which the nation-state emerged, and the role agricultural capitalism and the food trade played in that trajectory. It built on the insight that the articulated national economy, absent from Britain's workshop-of-the-world strategy, emerged on the margin of that world in the settler states.

As we know, the concept of the food regime took on a life of its own, becoming either a vehicle for historical analysis, or a lightning rod for critiques of mechanistic, structural and homogenising approaches to the dynamics of agri-food systems. The problem, I believe, lay in the separation of the concept from its original purpose. I return to that purpose in the light of current global developments, in the belief that a theorised history of the present is useful in critical reflection on the salience of our working concepts and our observations.

We have seen the collapse of the stable trading of the second (or post-World War Two) food regime, in a world awash with unregulated agricultural surpluses, accounting for the attention paid in the Uruguay Round in the 1980s, to the development of new rules of international agricultural commodity trade.

If the second food regime expressed the rules of the American hegemonic project (i.e. Bretton Woods, and the export of managed food surpluses and agri-industrial technologies as a geo-political strategy), what were the politics of the first food regime? These are not immediately apparent. Certainly there was a political dynamic in the rise of the nation-state as an alternative world-economic organizing principle to colonialism. But rules? Where were they, and under what conditions did international commodity trade occur? Was this strictly a regime?

I propose to argue that it was a regime in the sense that commodity production and circulation depended on the unique role of gold as a world money. The rules were implicit in the functioning of the sterling/gold standard. How did it work?

According to Karl Polanyi (1957), British capitalists used their political power in the post-Napoleonic era to create markets for land, labour and money to facilitate machine production and its insatiable need for price-governed inputs and market outlets. Thus was born the ideological construct of the 'self-regulating market', as a market complex relatively untrammelled by government intervention. Projected globally, this apparent self-regulating market needed a universal commodity equivalent, gold, to facilitate global exchanges among national and imperial markets. Gold became the world money, standardised in sterling balances held by governments in London, and through which the City redistributed liquidity to make the world go around. In turn, central banks used monetary policy to adjust the value of the national currency to the standard of value set by sterling/gold as the world money. Constitutional governments arose to represent those interests affected financially by currency adjustments.

More concretely, the 1834 Poor Law Act and the Bank Act of 1844 together institutionalised the English labour market and the central banking system, thereby stabilising accumulation, while the repeal of the Corn Laws in 1846 stimulated cheap grain imports from the 'new world' to reduce the cost of wage foods. Together, these measures instituted the markets in labour, money and land that underpinned Britain's prodigious commercial expansion on a global scale. In turn, Britain and the European states gained access to a broadening array of tropical products from the colonies (such as sugar, tea, coffee, oils, cotton, jute, rubber) and temperate products (grains and meat) from the settler states of North America and Australasia, to fuel industrial capitalism.

While British measures to institute commodity markets subjected the nineteenth-century world to the dynamics of industrial capitalism, they also generated the famous protective cycle of market regulation across the world of constitutional states. In this movement lay the various national forms of regulation: land markets and agricultural trade generating agricultural tariffs (representing early food security politics); labour markets generating social democratic responses (in domestic labour legislation and early import-substitution industrial strategies); and money markets generating currency management to stabilise national economic relations. Polanyi (1957: 203) emphasized the 'constitutive importance of the currency in establishing the nation as the decisive economic and political unit of the time.' In short, the protective response was formative of the nation-state.

In theoretical terms, the nineteenth-century world economy was indeed ruled by world money in the form of the sterling/gold standard. This was a wage labour regime, where a universal standard of money, expressing commodity equivalences, established the rule of the law of value in subjecting other forms of labour (slave, family farming, share-cropping, peasant labour) to the competitive dynamics of the industrial wage labour system (McMichael 1991). In other words, under this regime, other forms of labour and national currencies expressed their value, respectively, through the wage form and gold. Of course historically, this global circuit of value depended on military force, the City of London's pivotal role in organizing sterling balances, and England's aggressive commercial apparatus (see Polanyi, 1957; McMichael, 1985; Ingham, 1994).

This regime is seen in operation in Harriet Friedmann's (1978) seminal analysis of simple commodity production. This article examined the social function of a single world price for wheat in bringing geographically separated regions of wheat farming into competitive relations with one another. Friedmann highlighted the theoretical paradox that family farm labour in the new world out-competed hired labour on capitalist farms in England and Germany. An alternative highlight is that this outcome revealed the global social process of lowering the reproduction cost of industrial wages. By dismantling its corn laws, and determining the price of industrial labour by the cost of globally-sourced wage foods through the sterling-gold standard, British capitalism imposed a competitive valorising logic on agricultural producers across this world.

The collapse of the British-centred world economy in the early twentieth century resulted from a cumulation of protective counter-movements against market rule, culminating in national and imperial conflict among European states. The gold standard disappeared. In the post-World War Two period, monetary and wage relations were instituted nationally. The Bretton Woods monetary system supported the Keynesian principle of national full employment, pre-empting a world money regime like the sterling-gold standard. Trade was subordinated to national developmentalist policies, and this was expressed in the political management of the dollar-gold system of fixed exchange rates. The point of having a dollar-gold standard, backed by IMF short-term loans, was to limit the prospect of nations engaging in competitive devaluations (Leyshon and Thrift, 1997: 279), the kind of competitive devaluation characteristic of Asia and Latin America in the late 1990s.

The American dollar could not replicate the role of the British pound as a true world money. The difference was this: under the sterling-gold standard, by directly linking wages to prices (of goods and currencies), labour would shoulder the major burden of adjustment when over-extended or uncompetitive firms were devalued. Under the Bretton Woods monetary regime, labour was protected, locking American macro-economic policy into continual credit creation to maintain national accumulation and the rising social wage. Some of this credit found its way offshore, via aid programmes, into the growing Eurodollar accounts which represented potential claims against the dollar's value. Under this politically managed monetary regime, the link between value and money was increasingly artificial. As is well known, rising claims in the Eurodollar markets eventually forced the US government to remove the dollar from parity with gold, leaving an unstable system of floating currency exchanges.

The current post-Bretton Woods era rests on a system of floating currencies with no single world currency (other than the paper dollar standard), requiring the political management of world markets. Since there is no world money, and therefore no standard of value, the wage form no longer governs the valorisation of other forms of labour through the market. Rather, other forms of labour are valorised directly through political or non-market mechanisms. Labour forces across the world are cheapened, and this in turn devalues the wage contract through mechanisms such as the 'race to the bottom' (see Brecher and Costello, 1994). The de-institutionalisation of the wage form means late-twentieth century capitalism is concerned less with the reproduction of wage labour, and more with the reproduction of money (see Hoogvelt, 1997).

This shift is evident in the crisis of developmentalism: the public rhetoric is about jobs, but the private reality is about financial dealing and the casualisation of labour. Export processing zones proliferate, using coercive labour relations where the value of the labour reimbursement bears little relation to the labour power expended. The more generalised this phenomenon, the greater the pressure on the institution of wage labour, in addition to the profoundly destabilising consequences of the speculative conditions of this era of 'financialisation' (see Arrighi, 1994).

Politicisation of global market rule

The political foundations of this financial regime solidified in the management of the 1980s debt crisis. Debt management empowered the multilateral financial institutions, especially the IMF, which joined with the World Bank in establishing a distinctive form of politicised market rule.

In place of the automatic, or nationally mediated, adjustment of currency values associated with a world money system, we now have an overtly politicised operation led by international monetary fundamentalism. It is politicised in two senses: first, structural adjustment loans and programmes are directed at debtor states (i.e. their citizens) and not the banks holding their debt, thereby expressing the new power of financial capital; and second, adjustments are not simply economic, they are profoundly political insofar as they reorganise state structures and policies according to neo-liberal dictates of private efficiency, thereby expressing a form of financial colonialism (McMichael, 1995).

The first condition for loan rescheduling laid down by the IMF is currency devaluation. Currency devaluation, imposed under Article VIII of the IMF Articles of Agreement (and so far applied to about 90 member states), represents both an attack on the national currency and, following Polanyi (1957), on national sovereignty. Devaluation compresses real earnings, as domestic prices of food staples, essential drugs, fuel and public services inflate. Governments are then constrained to pursue anti-inflationary programmes, shrinking the state through reduction of public expenditure (including social programmes) and de-indexing wages to promote 'liberalisation of the labour market'. These conditions not only reduce national policy choices, but standardise them as well (to the extent that they do standardise in practice). They also demand political independence of central banks, allowing the IMF to handle money creation, which means resuming dependence on foreign loans. In global terms, IMF conditionalities reduce the value of labour in hard currency, and adjust domestic prices upward in a process referred to as 'dollarisation' (Chossudovsky, 1997: 56-9).

The cumulative effect of individual country adjustment is a general global restructuring, whereby labour costs are ratcheted downwards through de-indexation and the elimination of social wage supports. With the realignment of domestic prices to reflect world prices, and the exposure of unprotected domestic labour forces to the depressive forces of a world labour market, market rule is instituted through the collaboration between multilateral power and domestic political and economic elites who profit from state privatisation schemes and loan rescheduling. In effect, this unrestrained money power - unrestrained because money is now more a political relation than a value relation - privileges multilateral financial institutions (and transnational corporations) and reconstitutes state power around the implementation of monetarist orthodoxy (see Marazzi, 1997: 107).

This project, emanating from the debt regime, repeats the attempt to install a self-regulating market, this time in place of nationally instituted markets. Market freedom today means a frontal assault on the institutions of the nation-state (or citizen state). However, it is not that states are being eliminated; rather they are being restructured from nation-states into 'global' states, that is, institutions geared to securing global circuits of money and commodities and governed by 'consumer citizens'.

Global regulation is realised through the relaxation of national controls via new regulations. But it is by no means stable - this much is obvious from the recent South Korean debacle. Since money is no longer governed by commodity values, but by speculative circuits of financial capital, global regulation includes the need to avoid financial collapse, by underwriting and rescheduling debt on a decidedly *ad hoc* basis (and with apparently increasing frequency). This was, and is, exemplified in the emergency financial bail-out of beleaguered Asia-Pacific national banking systems in late 1997, by IMF packages supplemented with northern financial assistance. The IMF's role is to extract financial adjustment in the assisted states in order to sustain global capital flows. It was argued at the time that until these Asian states accepted IMF disciplines, firms and speculators would continue to pressure their national currencies (see Glain and Stein, 1997).

The lesson here is that capital is really unable to regulate itself, unless it submits to the law of value through a world money commodity like gold. Short of that, we have powerful debt security or bond-rating agencies like Standard and Poor's and Moody's (with combined listings of US$3 trillion), and financial speculators, which privately regulate the disposition of investment funds and the value of national currencies, respectively, according to financial orthodoxy (Sassen, 1996: 15-16). Money, and its regulation, has assumed a distinctively global politics.

Globalisation and the construction of a free trade regime

The debt regime was a dress rehearsal for a full-scale globalisation project that emerged in the 1990s, targeting all states and national economies. The GATT Uruguay Round was meanwhile codifying a free trade regime, for implementation by the World Trade Organisation (WTO), which was established in 1995. Complementing and strengthening this regime is the secretive OECD-initiated Multilateral Agreement on Investment (MAI). In its draft form, the MAI institutionalized the rights of corporations (and financiers) as investors, and created a legal status equivalent to that of nation states with rights to sue governments for breach of contract/free trade, without granting reciprocal rights to governments to sue such investors for damages on behalf of their citizens.

The codification of free trade for agricultural products in the WTO follows the informal expansion of export processing zones (EPZs) for manufactured goods. The object of EPZs is to reduce barriers to the free movement of capital and goods across national borders, to contain labour organisation, and to encourage the relocation of manufacturing processes from north to south. Where the EPZ anticipates market rule in selected enclaves, a WTO regime would generalise market rule, subjecting agricultural producers to the discipline of the global sourcing strategies of agribusiness corporations. Building on the intensification of agri-exporting to service debt, it promotes the recomposition of world agriculture around the abstract principle of comparative advantage.

In practice, the WTO regime is unlikely to be fully realised and sustained. The greater the extent of its implementation, the more unstable the world financial system becomes, and the stronger the political resistance grows. At the present time the world agricultural market is a hybrid, precisely because of the politicised character of markets. Northern policy-makers proclaim market principles as the key to accumulation, but continue indirect intervention, a recent form being 'green protectionism'. Some states are more equal than others in adopting free market policies, as Australasians well know. Through a side agreement known as the 'Green Box', negotiated bilaterally between the US and the EU, direct payments to farmers such as 'set-aside' payments (which are designed to remove certain land from production) are exempted from the definition of farm subsidies. In 1995, the industrial countries collectively paid US$182 billion in farm subsidies, 'equivalent to 41 percent of the value of production' (Watkins, 1996a: 247). While other subsidies are limited by the 1996 US Farm Bill, and 20 percent of export subsidies are to be reduced, the structure of farm subsidies will remain intact (Watkins, 1996b).

Thus, the legacy of the second food regime is continuing subsidies for northern farmers to overproduce, with the GATT/WTO rather than the US PL-480 programme, now managing the disposal of food surpluses by requiring that southern states liberalise their farm sectors. The Uruguay Round stipulated that:

> All but the least developed countries will be required to reduce their tariffs on food imports by 24 percent over ten years and to increase the minimum level of imports from one percent to four percent of consumption (Watkins, 1996a: 251).

Such developments can be illustrated by events in the early 1990s, when the US Department of Agriculture (USDA) estimated that by the year 2000, Southeast Asia would account for two-thirds of the US$3 billion-plus increase in global demand for farm exports, assisted by over US$1 billion in US Export Enhancement Program subsidies to exporters. A portion of this lucrative market (much of which consists of tinned beef and processed foods sold to South Korea and Taiwan) would involve bulk wheat and corn imports by Indonesia, Malaysia and the Philippines. The USDA (cited in Watkins, 1996a: 253) predicted that:

> In the absence of sustained, aggressive investment in infrastructure and increased competitiveness for corn production, the Philippines could become a regular corn importer by the end of the decade...US corn may be able to capture a large share of this growing market

Under the conditions of the 1994 agricultural agreement of the Uruguay Round, imported US corn, predicted by OECD projections to undercut corn prices in the Philippines by 20 percent in the year 2000, would depress domestic corn prices, threatening half a million peasant households with income declines of 15 percent (Watkins, 1996b). This would involve high social costs, such as reduced expenditure on education, increased reliance on child labour, nutritional decline, and the intensification of women's work outside the home to compensate for these effects. Comparatively, the average subsidy to US farmers and grain traders is roughly 100 times the income of a corn farmer in Mindanao. As Kevin Watkins (1996b) remarks,

> In the real world, as distinct from the imaginary one inhabited by free traders, survival in agricultural markets depends less on comparative advantage than upon comparative access to subsidies.

Watkins (1996b) notes that the government of the Philippines views this agreement as immutable and as an instrument of economic efficiency which effectively transfers sovereignty over national food policy from Manila to a remote and unaccountable trade body in Geneva, and concludes:

> Legal niceties aside, the Uruguay Round agreement bears all the hallmarks of an elaborate act of fraud. It requires developing countries to open their food markets in the name of free market principles, while allowing the US and the EU to protect their farm systems and subsidise exports.

Free trade agreements like NAFTA mirror the asymmetry of the Uruguay Round agreement. Quotas on duty-free US corn, wheat and rice imports into Mexico are being lowered in stages. In Mexico, 2.5 million households engage in rainfed maize production, with a productivity differential of 2-3 tons per hectare compared with 7.5 tons per hectare in the American mid-West. With an estimate of a 200 percent rise in corn imports when NAFTA is fully implementated in 2008, it is expected that more than two-thirds of Mexican corn production will not survive the competition - especially as the 1994 financial crisis has undercut the Mexican state's transitional income support programme (Watkins, 1996a: 253).

Stepping back, we note that the US controls over 75 percent of the world market for corn, and 80 percent of the market for soya and sorghum, allowing the US and its grain traders virtually to dictate world prices. The infrastructure of US 'green power' allows traders to purchase commodities at artificially low prices and then, with the help of export subsidies, dump them on the world market - at around half the price guaranteed to farmers (Watkins, 1996b). In other words, commodity prices are systematically delinked from production costs. Such global political management of markets recomposes food production and consumption relations on a broad scale.

What is often forgotten in the liberalisation rhetoric is that the free world market does not simply realise the 'comparative advantage' of extant efficient producers across the world, but that its powerful agents also recompose what is produced and where. For example, food companies like Cargill and Continental, which account for 50 percent of grain exports from the US, 'now have the power to shift comparative advantage simply by their decisions on where to build warehouse, transport, and processing facilities' (Lehman and Krebs, 1996: 125).

Conditions of agri-food system recomposition

The recomposition of agricultures has accelerated under the combined forces of structural adjustment, transnational corporate strategies such as global sourcing (with flexible subcontracting networks) and genetic patenting, and what we can term the 'import complex'. Each is a feature of global restructuring, and expresses the current form of politicised market rule. Each is considered via a number of case studies.

Structural adjustment

The conditions of structural adjustment imposed on Somalia in the 1980s resulted in the collapse of agricultural infrastructure and an 85 percent decline in public expenditure on agriculture. Currency devaluation pushed up prices of farm inputs. The multilateral donors provided aid in the form of cheap food to be sold by the government, which in turn generated counterpart funds as the main source of state revenue, 'thereby enabling donors to take control of the entire budgetary process' (Chossudovsky, 1997: 104). Between 1975 and 1985, food aid increased 15-fold at an annual rate of 31 percent, helping to shift diets away from staples such as maize and sorghum, towards wheat and rice. The resulting crisis for local farmers was an opportunity for bureaucrats, military elites and merchants with political connections to appropriate the best agricultural lands and develop 'high value-added' fruits, vegetables, oilseeds and cotton for export (Chossudovsky, 1997: 102). A similar pattern has occurred across Africa, with grain imports to sub-Saharan Africa rising from 3.71 million tons in 1974 to 8.47 million tons in 1993, and cheap beef and dairy products imported from the European Union increasing 7-fold since 1984, and overwhelming the West African pastoral economy (Chossudovsky, 1997: 106).

Transnational corporate strategies

The corporate strategies of transnational companies (TNCs) build on the above conditions. One of the Mexican government's gestures for membership of NAFTA was to revise Article 27 of the revolutionary constitution to allow the private alienation of collective *ejido* lands, and to add a clause allowing 100 percent foreign ownership of firms. In the context of these legal revisions, US, European and Japanese corporations have invested extensively in fruit and vegetable production, which accounts for about 40 percent of Mexico's agricultural exports, but only 6 percent of its cultivated land (Watkins, 1996a: 251). In Chile, currently the largest supplier of off-season fruits and vegetables to Europe and North America, more than 50 percent of fruit exports are controlled by five TNCs. Not only have these 'non-traditional' agri-exports recomposed local agricultural landscapes, but they have also casualised agricultural labour - roughly two-thirds of the Mexican and Chilean agricultural labour forces depend on insecure, low wage employment. In Mexico, 'workers on commercial farms are generally paid piece rates, are not entitled even to minimal social welfare protection such as sickness and maternity benefits and have no basic trade union rights' (Watkins, 1996a: 251).

The import complex

The concept of the 'import complex' describes a new organisational logic of the global political economy under market rule. It is expressed in the well-known distributional statistic that roughly 80 percent of the world income is produced and consumed by 15 percent of the world's population. The other side of this coin is that 80 percent of the world's six billion people do not have access to consumer cash or credit - in other words, the global economy we are talking about is global in relational terms, but not in social or geographical terms.

In the global economy, much of the material production occurs in select regions of the south, to be consumed largely in the north, although this is a relational rather than an absolute statement, as there are certainly southern-regional nodes of affluent consumption. This relational complex has a number of historical, political and technological conditions, not the least of which is the rarely discussed energy subsidy to the corporations which transport components and finished products across the world, and which constitutes part of the huge comparative advantage to TNCs and northern consumers. An increasingly consequential condition of this relational complex is the construction of a global labour reserve which, through the mechanisms already discussed, generates a global political-economy based on cheap labour.

The post-industrial north has been characterised as a 'rentier economy' which appropriates wealth in both material and value terms from parts of the south (Chossudovsky, 1997: 85). Through monopoly control of the services sector, high technology firms control value-added in manufacturing and agricultural production, and appropriate the profits associated with commodity circulation. The so-called NIKE pyramid (including the statistic that basketball player Michael Jordan's annual royalties of US$20 million from NIKE exceed the combined incomes of that company's 75,000 Asian employees) is one indicator of this. Another is the fact that the retail price of coffee in OECD countries is 7 to 10 times the recorded import (f.o.b.) price, and 20 times the price paid to the southern farmer. In the garment trade, a Paris-designed shirt purchased for US$4 in Bangladesh or Thailand by a global fashion designer will sell at a price of US$20 to US$40 in the European market. Thus the 'GDP of the importing Western country increases without any material production taking- place' (Chossudovsky, 1997: 87). Although these are global relations, they intensify through the construction of regional free trade agreements, which lock these political-economic structures in place. Through NAFTA, for instance, American firms reduce their labour costs by more than 80 percent through relocation of plants to, or subcontracting in, Mexico, where wages are up to 10 times lower (Chossudovsky, 1997: 83).

In the twenty-five years up to 1997, the rise of the East Asian import complex has accompanied the rise of East Asian capitalism as a growth pole of the global economy (McMichael, forthcoming). We can relate the evolution of the East Asian food import complex to the rise and demise of the postwar food regime via three phases. During the first, early postwar phase, the principal food import into Japan was wheat (supplemented with some rice imports), under the American food aid programme. From the 1960s on, the expansion of animal protein consumption supplemented, and to some extent displaced, wheat consumption, with rising imports of feedstuffs and controlled beef imports. The second phase (from the 1970s on) included joint public-private offshore ventures sponsored by the Japanese government's food-security driven import-diversification strategy, which dovetailed with the emergence of the 'new agricultural countries' (NACS) such as Brazil, Mexico, Chile and Thailand as agri-export platforms. The third phase coincides with the increased capital mobility of the 1980s when, under the combined forces of debt servicing and liberalisation pressures, a new flexible form of global sourcing emerged at the initiative of transnational firms, especially food companies.

While the first two phases serviced government-managed shifts in national consumption patterns, the latter phase extends the initiative and scope to the world market and to food corporations. In particular, this phase involves a qualitative recomposition of regional agricultures via incorporation into selective global circuits (as well as the displacement of Japanese farmers). The agents of this new dynamic are transnational firms, which lobby governments and international institutions like the GATT/WTO.

The recomposition of regional agricultures includes agri-exporting to service debt, and the offshore movement of Japanese agribusiness capital seeking to reduce costs, exacerbated by *endaka*, and anticipating liberalisation of Japanese food imports (Riethmuller, 1992). The Japanese food processing industry, for example, expanded its foreign investments fourteen-fold between 1985-1989 (Yamauchi and Nakashima 1996). Increased imports of agricultural products, to reduce domestic food prices, was recommended by the Maekawa Report commissioned by Prime Minister Nakasone in 1986 (ABARE, 1988; Ohno, 1988: 22). As Riethmuller (1992: 31) reports,

> While imports of rice and wheat are not allowed unless official permission has been given, imports of processed foods containing rice and wheat are permitted - such as rice pilaf, rice cake cubes, rice powder, and flour products such as biscuits, noodles and pasta, in addition to frozen vegetables and processed pork.

The fact this process is partly policy-driven relates the import complex to what Gavan McCormack (1996) terms the Japanese 'consumer state'. The concept of the 'consumer state' is an unstable and unsustainable feature of late-twentieth century global capitalism, characterised by speculatively driven over-production (in the context of the cheapening of the global labour force). Domestically, northern 'consumer states' are increasingly 'two-third societies', where one-third of the electorate is marginalised by long-term unemployment (Hoogvelt, 1997: 147). Consumer states depend increasingly on the 'politics of exclusion', packaged in the discourse of social capital and the privatisation of security and services, all designed to transform the 'citizen state' by reconstructing citizenship along neo-liberal lines (Drainville, 1994; Hoogvelt, 1997: 149). This devaluation of the 'citizen state' is a destabilising process, especially in the context of increasingly frequent (global) financial crises.

Internationally, the affluent consumer state leaves huge ecological and dietary footprints. For example, in the Philippines, 55 percent of farm land is devoted to export crops, including bananas and pineapples for Japanese consumers; in Thailand, between 1961-1989, mangrove forests declined by half to fuel the 'blue revolution' in shrimp farming for export to Japan (McCormack, 1996: 133); and the conversion of the Australian beef industry from extensive grazing into intensive lot-feeding for Japanese markets has severely compromised the local agricultural ecology (Lawrence and Vanclay, 1994). The deepening of these footprints expresses the unsustainability of the 'import complex' dynamic.

Conclusions

The relations underlying the 'import complex' suggest a dynamic akin to the myth of Midas' touch, where the King's gold fetish was, it may be said, ontologically unsustainable. In my view, indiscriminate valorisation is generating a multiplicity of contradictions across a world that is selectively plundered for its human, mineral, biological and genetic resources.

The generation of wealth is no longer linked to surplus value extraction through the wage relation alone. Much wealth is artificially created by credit, accelerating the concentration and centralisation of capital and its technological base (devaluing wage labour), and intensifying the consumption of productive and increasingly non-productive (e.g. symbolic) commodities. Neo-liberal, supply-side economics has resulted in a crisis of overproduction, as consumption has not matched the expanded global production base - e.g. the global auto industry now has the capacity to produce 80 million vehicles a year, for a marketplace with fewer than 60 million buyers. Such overcapacity puts pressure on firms to lay off workers and relocate to reduce their labour costs, further exacerbating the problem (Greider, 1997).

The augmentation of credit relations stems from the hyper-mobility of speculative capital, aided by computer and telecommunication technologies, and the decoupling of financial from productive capital (Arrighi, 1994). Beyond credit there are the futures markets, driving the process of monopolisation of capital and of future resources.

In all these senses, it can be argued that the present is no longer the logical development of the past, rather it is increasingly the hostage of the future. Certainly those who promote the project of globalisation would have us believe that the train is leaving the station and unless we get on board we will forfeit the benefits of a truly global economic system. Also, it seems clear enough that, in the light of escalating global environmental and health concerns, the social psychology of the forthcoming millenium will include a nagging fear that humans may join the list of endangered species (see McMichael, 1993).

The challenge of a global perspective such as this is its credibility. The world is far more complex and messy than the theoretically-driven characterisation that I have laid out here. Some argue that a global view imposes a singular, or categorical, logic on a geographically and culturally diverse world (see for example, Whatmore, 1994). But this is exactly what I am arguing. And besides, social science, at whatever scale, applies a categorical logic to its chosen topic, whether it is social relations or a sample of interviewees. Categories and concepts organise our understanding of reality. If we are careful, they express historical relations and do not become static, ideal-typical or trans-historical. This is my goal here - to critically interpret the present by situating it in comparative-historical perspective. Abstraction is essential to the historical method. The categories I have chosen to employ are historical and relational categories that suggest ways in which global processes are embedded in local processes.

Let me illustrate what I mean with a re-interpretation of some recent phenomena associated with agri-food restructuring, such as the revival of sharecropping, the spread of contract farming, and the increased employment of women in agriculture. In an historical conjuncture, where the wage form governs value production and exchange less and less, non-wage forms of labour become at once increasingly significant and increasingly tenuous. Included in non-wage forms is the semi-wage, or temporary hired worker. Jane Collins (1995) has documented recent trends towards the feminisation of Latin American agricultural workforces. Agribusiness firms hire women to combine high-quality labour with the lower costs associated with the flexible employment patterns of women, which is related to their primary responsibility to provision their household - in other words, capitalist social relations are not simply market relations, but implicate household relations also as part of their conditions of reproduction. Collins (1995: 217) concludes:

> Agribusinesses use gender ideologies to erode stable employment and workers rights where women are concerned. Of equal significance, employing women provides the employer with a way of invoking institutions beyond the workplace to extend and reinforce labour disciplines.

My point is that in a world market where valorisation has to be managed politically, all kinds of social and cultural conditions beyond the formal labour market, enter into the construction of labour forces vulnerable to human and labour rights abuses. In Chile, the showcase of structural adjustment, the Labor Code formally resembles the informal practices in the US labour market (which appear to be the latest American export). That is, it allows:

> employers to fire workers at will, individually or en masse, for 'business' necessities,' eliminating, according to the architects of the code, the 'monopoly' many workers had on their jobs....the right to organize was extended only to workers employed for at least six consecutive months (Watkins, 1996a: 160).

This includes the Chilean fruit industry where women work a few months a year, on a piece rate basis - a standard feature of a labour market where the wage relation is not regulated by the law of value.

In her book *Strawberry Fields*, Miriam Wells (1996) documents the revival of sharecropping in California, arguing that its use was a class strategy on the part of growers to undercut the power of organised farm labour. In Wells' (1996: 302) terms, not only are sharecroppers 'essentially employees with a share feature to their wage contracts', but also they use labour contractors to hire devalued labour:

> Research shows that the use of contractors lowers wages and benefit levels, impedes labour organising, increases worker dependency, and reduces the likelihood that employees will pursue their rights under the law (1996: 299).

The historical context is not simply the decline of the wage contract, but the increased regional articulation of this form of labour with the Mexican subsistence, or informal, sector, which subsidises the reproduction of severely underpaid and underemployed labourers on sharecropper plots (1996: 285).

The third example is drawn from Little and Watts' (1994) volume on the intensification of contract farming in sub-Saharan Africa. In his introductory chapter, Watts (1994: 23-4) notes that contract farming expresses the extension of agri-industrial production to peasant cultures as the 'new agriculture' alternative to traditional plantation systems, thereby transforming rural life. In discussing the conversion of the free peasant into an unfree grower, Watts (1994: 71) concludes that,

> Contract farming illuminates new configurations of state, capital, and small-scale commodity production within a changing international division of labour, thereby challenging the essentialist views of agrarian capitalism and labour-capital relations more generally.

In these cases, the analysts discover that contemporary agribusiness operations increasingly depend on non-wage forms of labour. The absence of the linear development of capitalist wage labour forms implies some contradictory, or politically mediated, relations between the state, firms and labour, within a global economy. My observation is that such diversity of agricultural labour forms expresses something more than simply the flexibility of capitalism in its pursuit of lower labour costs, resources and perhaps quality control. These phenomena are historically specific. Debates about whether these labour forms are 'disguised wage labour' may no longer be the point, if we understand wage labour to have defined an historical period in the evolution of capitalism (such as I have suggested). That is, the development of wage labour is neither a secular trend, nor perhaps will it remain the defining feature of capitalism. Arguably, the wage labour regime was more a world-historical moment than the empirical basis for a benchmark concept. The proliferation of non-wage, or casual-wage, forms of labour today perhaps defines a historical period in which the wage labour contract is eroding, and in which that erosion is both a condition of, and is conditioned by, the attempt to elaborate a politicised form of market rule on a global scale.

Note

1   I wish to acknowledge the helpful comments of Dale Tomich and Rajeev Patel on earlier drafts of this paper.

References

Australian Bureau of Agricultural and Resource Economics (ABARE), (1988) *Japanese Agricultural Policies: A Time of Change*, Australian Government Publishing Service: Canberra.

Arrighi, G. (1994), *The Long Twentieth Century: Money, Power and the Origins of Our Times*, Verso: London.

Brecher, F. and Costello, T. (1994), *Global Village or Global Pillage? Economic Reconstruction from the Bottom Up*, South End Press: Boston.

Chossudovsky, M. (1997), *The Globalisation of Poverty: Impacts of the IMF and World Bank Reforms*, Third World Network: Penang.

Collins, J. (1995), 'Gender and Cheap Labor in Agriculture', in McMichael, P. (ed), *Food and Agrarian Orders in the World Economy*, Praeger: London, pp. 217-232.

Drainville, A. (1994), 'International Political Economy in the Age of Open Marxism', *Review of International Political Economy*, Vol. 1, No. 1, pp. 105-132.

Friedmann, H. (1978), 'World Market, State and Family Farm: Social Bases of Household Production in an Era of Wage Labor', *Comparative Studies in Society and History*, Vol. 20, No. 4, pp. 545-86.

Friedmann, H. (1987), 'Family Farms and International Food Regimes', in Shanin, T. (ed), *Peasants and Peasant Societies*, Blackwell: Oxford.

Friedmann, H. and McMichael, P. (1979), 'Agriculture and the State System: The Rise and Decline of National Agricultures', *Sociologia Ruralis*, Vol. 29, pp. 93-117.

Glain, S. and Stein, P. (1997), 'Worries Remain After Asian Markets Stage a Rebound', *Wall Street Journal*, 30 October, p. A19.

Greider, W. (1997), *One World, Ready or Not: The Manic Logic of Global Capitalism*, Simon and Schuster: New York.

Hoogvelt, A. (1996), *Globalization and the Postcolonial World: The New Political Economy of Development*, Macmillan: London.

Ingham, G. (1994), 'States and Markets in the Production of World Money: Sterling and the Dollar', in Corbridge, S., Martin, R. and Thrift, N. (eds), *Money, Power and Space*, Blackwell: Oxford, pp. 29-48.

Korten, D. (1995), *When Corporations Rule the World*, Kumarian: New York.

Lawrence, G. and Vanclay, F. (1994), 'Agricultural Change and Environmental Degradation in the Semi-periphery: The Case of the Murray-Darling Basin, Australia', in McMichael P. (ed), *The Global Restructuring of Agro-Food Systems*, Cornell University Press: Ithaca, pp. 76-103.

Lehman, K. and Krebs, A. (1996), 'Control of the World's Food Supply', in Mander, J. and Goldsmith, E. (eds), *The Case Against the Global Economy and For a Turn Toward the Local*, Sierra Club Books: San Francisco, pp. 122-130.

Leyshon, A. and Thrift, N. (1997), *Money Space: Geographies of Monetary Transformation*, Routledge: London.

Lietaer, B. (1997), 'From the real economy to the speculative', *International Forum on Globalization News*, Vol. 2, pp. 7-10.

Marazzi, C. (1997), 'Money in the world crisis: the new basis of capitalist power', *Zerowork*, Vol. 2, pp. 91-112.

McCormack, G. (1996), *The Emptiness of Japanese Affluence*, M. E. Sharpe: Armonk, NY.

McMichael, A. (1993), *Planetary Overload: Global Environmental Change and the Health of the Human Species*, Cambridge: London.

McMichael, P. (1985), 'Britain's Hegemony in the Nineteenth-century World Economy', in Evans, P., Rueschemeyer, D. and Stephens, E. (eds), *States Versus Markets in the World System*, Sage: Beverley Hills, pp. 117-150.

McMichael, P. (1991), 'Slavery in the Regime of Wage-labor: Beyond Paternalism in the U.S. Cotton Culture', *Social Concept*, Vol. 6, No. 1, pp. 10-28.

McMichael, P. (1995), 'The 'New Colonialism': Global Regulation and the Restructuring of the Inter-state System', in Smith, D. and Borocz, J. (eds), *A New World Order? Global Transformations in the Late Twentieth Century*, Greenwood Press: Westport, pp. 37-55.

McMichael, P. (forthcoming), 'Global and Regional Implications of the East Asian Food Import Complex', *World Development*.

Ohno, K. (1988), 'Japanese Agriculture Today - Decaying at the Roots', *AMPO Japan-Asia Quarterly Review*, Vol. 20, No. 1/2, pp. 14-28.

Polanyi, K. (1957), *The Great Transformation: The Political and Economic Origins of Our Times*, Beacon: Boston.

Riethmuller, P. (1992), 'Japanese Direct Foreign Investment in Agricultural Industries: A Review of Recent Developments', *Agribusiness*, Vol. 8, No. 1, pp. 23-33.

Sassen, S. (1996), *Losing Control? Sovereignty in an Age of Globalization*, Columbia University Press: New York.

Wallerstein, I. (1983), *Historical Capitalism*, Verso: London.

Watkins, K. (1996a), 'Free Trade and Farm Fallacies: From the Uruguay Round to the World Food Summit', *The Ecologist*, Vol. 26, No. 6, pp. 244-255.

Watkins, K. (1996b), *Trade Liberalisation as a Threat to Livelihoods: The Corn Sector in the Philippines*, at <http://www.oneworld.org/oxfam/policy/research/ corn.htm>

Watts, M. (1994), 'Life Under Contract: Contract Farming, Agrarian Restructuring and Flexible Accumulation', in Little, P. and Watts, M. (eds), *Living Under Contract: Contract Farming and Agrarian Transformation in Sub-Saharan Africa*, Wisconsin University Press: Madison, pp. 21-77.

Wells, M. (1996), *Strawberry Fields: Politics, Class, and Work in California Agriculture*, Cornell University Press: Ithaca.

Whatmore. S. (1994), 'GIobal Agro-food Complexes and the Refashioning of Rural Europe', in Amin A. and Thrift, N. (eds), *Globalization, Institutions and Regional Development in Europe*, Oxford University Press: Oxford, 46-67.

Yamauchi, H. and Nakashima, Y. (1996), *The Japanese Food Industry System*, at <http://www2.hawaii.edu/apfat/iep1.htm>.

2 Switzerland's billabong? Brand management in the global food system and Nestlé Australia

BILL PRITCHARD

Introduction

Since the late 1980s, the broad sweep of political economy research into the globalisation of agri-food systems has been weighted heavily towards issues of production, trade and raw materials sourcing. No doubt this focus is a response to the desire to understand and document the considerable restructuring of these processes that has occurred through the 1980s and 1990s. Yet this focus has also, perhaps unwittingly, drawn attention away from less tangible manifestations of agri-food globalisation. In particular, there has been a dearth of research relating to issues associated with the strategic roles of intangible assets, including most obviously the global spread of product and corporate branding.

This paper presents a preliminary excursion into questions relating to the role of brand names within processes of agri-food globalisation. It undertakes three tasks. First, it establishes a framework for examining the roles of brand names within agri-food globalisation. Second, it situates this framework within the corporate strategy of the Swiss-owned Nestlé S.A., the world's largest food company. Third, it applies these concepts through a case study of the branding strategy of Nestlé Australia subsequent to its 1995 takeover of the Australian ice cream manufacturer, Peters.

Brand names in the global food system

The McDonald's brand is the beacon that guides our future as surely as it illuminates our present and reflects our past. Our brand is more than a road sign, more than our Golden Arches logo, more than the Big Maċ or speedy service - it is the sum of the entire McDonald's experience. Our brand lives and grows where it counts the most - in the hearts of our customers (McDonald's Corporation, 1996: 27).

The above quote above is taken from the 1996 Annual Report of the McDonald's Corporation. Although the tone of this quote is undeniably lyrical, it nonetheless underscores an important financial characteristic of the global food industry: namely, that brand names and other forms of intangible assets can comprise major components of the net asset worth of large food companies. Furthermore, the development, nurturing and acquisition of brand names and other intangible assets can be the anchors of corporate strategies, and thus impose a major influence on the trajectories of individual companies and industries. According to 1994 calculations, intangible assets comprised 86 percent of the net asset worth of Danone (a French-owned transnational food company), and 78 percent of the net asset worth of Grand Metropolitan (a British-owned transnational food company) (Ball, 1994).

The strategic importance of intangible assets is widely acknowledged by analysts in equities and securities markets, and by commentators in the business and trade media. Specialist international consultancy firms provide brand strategies and offer brand valuations. One of the largest of these consultancy firms, Interbrand, defines its corporate philosophy in the following terms:

> For much of the twentieth century, the vast majority of a company's assets were tangible - real estate, plant, facilities, equipment, inventory, stock investments and cash. A company's balance sheet was, therefore, a reliable measure of its worth. However, this is no longer the case. Today, the intangible assets of the firm are frequently the most valuable - that is, the copyrights and patents; the company's technical expertise and know-how; the management teams, and particularly, the company's trademarks or brands. We believe that brands are now among a company's most valuable assets and represent the 'engines' of corporate growth, future success and ongoing profitability. Given this situation, it is imperative that companies understand the quality of these important assets and manage them for distinctiveness and growth (Interbrand, 1997).

The concept that intangible assets can be the focus of corporate strategy resonates deeply in the food industry. For example, the global strategy of the Sara Lee Corporation, which in 1996 was the world's seventeenth largest food company, is defined strictly in terms of brand ownership. Sara Lee's 1996 Annual Report defines the company's objectives in the following terms:

'We are not a food company or an apparel company or a shoe company. We are - and shall continue to be - a global marketer of branded consumer packaged goods' (Sara Lee Corporation, 1996: 3). In September 1997, the company entrenched this philosophy with a US$1.6 billion corporate restructuring involving the sale and outsourcing of production operations (*The Economist,* 1997a: 81). This meant that whereas Sara Lee continued to own its brands, it was increasingly the case that it did not own the facilities to manufacture the products attached to those brands.

Somewhat surprisingly, research into globalisation of the agri-food system in general has given marginal attention to the politico-economic context in which intangible assets are developed, and to which they relate. The role played by intangible assets is often overlooked in analyses of restructuring. There is a general tendency for political economists and economic geographers to fix their gaze on the roles of physical rather than intangible assets, when examining restructuring processes. Corbridge and Thrift (1994: 2) argue that 'rather more of the work conducted on the two 'posts' (post-Fordism and postmodernism) has been about visible fixed points and patterns in production than about the invisible spatial flows that link these nodes together'. To the extent that these issues have attracted a research focus, this has come from analyses of retail restructuring, in particular the roles of own label brands and their impacts on the relative power balance between retailers and food manufacturers (Hughes, 1996; Sparks, 1997).

It is possible to speculate about the reasons for the general neglect of these issues. It may be because geographers and political economists, with their training for fieldwork, have a predilection for studying physical manifestations of economic activity (as embodied in factories, farms and retail centres). It may be also because the examination of intangible assets may appear to presuppose knowledge of corporate accounting, for which many geographers and political economists are untrained. These biases and impediments, however, should not overshadow the need to understand the processes by which companies use brand strategies within their global objectives.

The ways companies create, develop and exploit brand names are linked to a wider set of processes in the contemporary capitalist space-economy. Patterns of corporate restructuring since the mid-1970s are linked inextricably to a heightened international mobility of capital (Bryan, 1995: 9-25). Most transnational corporations now undertake extensive international bench-marking exercises based on rates of return criteria. Companies shift their asset holdings with relative impunity, as they search for globally optimal portfolio returns. This type of corporate management strategy relates to, and is expedited by, the ability to shift finance capital across international borders. On the one hand, it relates to the establishment of an international market for investment, via globally mobile superannuation/pension funds. Fund managers shift their share holdings according to these expectations of international rates of return. On the other hand, the global mobility of finance capital has helped create an international market for corporate control, through mergers, acquisitions and joint ventures. This environment has been underpinned by a series of deliberate state interventions aimed at increasing the international mobility of capital (including the abolition of exchange controls, the liberalisation of international banking systems, and the establishment of bilateral taxation treaties) and introducing technological changes that have, in distinctive ways, 'shrunk the world'.

In general, these processes amplify the importance of branding and corporate image within profit accumulation strategies. They have sharpened processes of capital accumulation within individual economies, so that the overall pace of economic restructuring accelerates. In key respects, these changes represent a continuation of themes and trends dominant for most of the twentieth century (Hirst and Thompson, 1996). The global mobility of capital, although impressive, is not totalising (Whatmore and Thorne, 1997). Nonetheless, it seems clear that the heightened global mobility of capital over the past two decades has also produced qualitative changes to the global economy (Tickell and Peck, 1992). There is a certain 'newness' in the operation of capitalism that cannot easily be ignored. The central element of this 'newness' relates to the relationship between (highly mobile) finance capital, and (relatively more spatially fixed) production capital. Increasingly, this has meant that corporate strategy requires the formulation of complex intra-firm relationships between product sourcing, brand ownership and marketing. The case of Nike Inc. provides a quintessential example of how, through the development of a complex sub-contracting network in East Asia, brand ownership and marketing can be separated from production (Donaghu and Barrf, 1990).

A second element of the global economy's 'newness' impacting on corporate strategy is the way that the increased global mobility of finance flattens key aspects of geographical distance and difference. Since the mid-1980s, the key arena of international corporate lending has been the international bond market (Bryan, 1995: 17), which operates outside the direct domain of central banks. Roberts (1994) labels these markets as operating in 'fictitious spaces'. These markets trade derivatives, such as interest rate swaps, which exist 'only in conceptual terms' (Leyshon, 1996: 70). Companies use these products in an endeavour to obliterate the geography of financial access and risk that confronts them in their particular spatial environments. Their existence helps establish new arenas of financial risk, access and exclusion (Corbridge and Thrift, 1994: 15-22). These arenas can exist at odds with 'taken-for-granted' notions of national financial and economic space.

This environment facilitates the release of intangible assets, such as brand names, into an international arena of trade and exchange. This process is seen most readily in the case of international mergers and takeovers, where acquisition strategies are often built on the capture (and sometimes resale) of intangible assets such as branding rights or trademarks. In the 1980s, a number of takeover strategies were constructed around accounting rules which allowed intangible assets to be revalued, and thus offered immediate 'paper' profits for successful acquirers. This was certainly a feature of the protracted takeover battles between the Australian-New Zealand transnational Goodman Fielder Wattie, and the British transnational Rank Hovis McDougall PLC, during 1988-89 (Napier, 1994: 76). Moreover, there remain certain flexibilities in the ways intangible assets are valued. Many brands exist in a commercial climate that is extremely dynamic. Their net worth can depend on a host of factors inside and outside a company's control.

The capacity for intangible assets to be traded internationally opens new 'possible worlds' for companies and corporate strategy. These worlds are less strictly defined by the economic and financial boundaries of national spaces (Leyshon, 1996: 77). As a result, the 'production geographies' of companies can be sharply askew to their 'finance geographies', including the places at which intangible assets (such as brand names) are held for legal purposes. Tax havens and global financial centres have assumed a role of 'niche political geographies' for finance capital, and have become increasingly important as finance has become more mobile (Connell and Pritchard, 1990). For companies, their existence creates new geographies of money and credit, which can drive the ways in which productive activities are financed, and in which profits and royalties are repatriated.

One of the clearest, though least researched, elements of this process is the way intangible assets are sited and traded. Intangible assets such as brand names carry rights, which can generate royalty streams. Companies can possess considerable freedoms in nominating the ownership structures of intangible assets. Hypothetically, this may include intangible assets being registered in a wholly owned corporate subsidiary located in a tax haven (or at least, a country with more favourable taxation rules). These ownership structures can influence the shape, direction and size of international royalty flows. These flows have implications for profit rates and taxable incomes in specific political arenas. National tax authorities have different guidelines regarding the tax treatment of brands. For example, the Dutch and Irish tax authorities allow companies to depreciate the value of their brands for tax purposes, unlike the stance taken by these authorities in Britain (Ball, 1994: 40).

Where royalty payments exist as related-party transactions (that is, royalties are paid by a corporate subsidiary to its corporate parent), there may exist considerable potential for rescheduling royalty and profit rates in line with opportunities for international tax minimisation. This is a 'grey' area of corporate and taxation law which is difficult to police. Often, intra-firm royalty transactions are connected to intra-firm trade in tangible goods, as companies coordinate production and sales strategies internationally. As a result, taxation authorities may be required to undertake lengthy and expensive audits in order to ensure companies' related party transactions meet legal guidelines. In a recent Australian taxation ruling, the Commissioner of Taxation acknowledges the ambiguities inherent in this exercise:

> For example, a lump sum payment made under an agreement providing both for the outright sale of machinery used to produce certain articles and for the right to manufacture and sell those articles under a brand name would be a royalty to the extent that it was attributable to the right to use the brand name. No hard and fast rules can be applied in determining how much of a payment represents a royalty. Each situation has to be resolved on its own particular facts (Australia, Commissioner of Taxation, 1997a: paragraph 13).

The United States Inland Revenue Service (IRS) has been at the forefront of endeavours to monitor related party transactions of transnational corporations. In 1991 the IRS introduced advance pricing agreements (a process by which transnational corporations could negotiate in advance with tax authorities regarding related party transactions), and in 1992 introduced extensive documentation requirements for related party transactions. In 1995, the OECD issued revised transfer pricing guidelines aiming to take account of the increasingly complex nature of global intra-firm trade.

In Australia's case, the establishment of a related-party international royalty payment (say, from local subsidiary to its foreign brand-owning parent) may provide opportunities for companies to reduce their Australian corporate taxation liabilities. Royalties paid offshore to a parent attract a withholding tax of 15 per cent, which is considerably below the statutory company tax rate of 36 per cent. However, companies must comply with various provisions of the Income Tax Assessment Act and Rulings of the Tax Commissioner, when structuring their related-party royalty payments. The Income Tax Assessment Act (division 13) specifies that offshore related-party royalty payments be structured as if they were arms-length transactions. Moreover, companies liable to make such royalty payments must first gain approval from the Commissioner of Taxation of the amount of tax to be deducted (Australia, Commissioner of Taxation, 1997b: paragraph 7).

The task of assigning values to related-party payments as if they were arms-length transactions often can provoke dispute between taxation authorities and companies. As a result, transnational corporations in recent years have tended to give considerable attention to issues relating to their management of related-party transactions. A 1997 survey of 393 transnational parent companies and 76 transnational subsidiaries by the accounting company Ernst and Young discovered that transfer pricing was the major tax issue as perceived by transnational corporations. Furthermore, the survey found that the tax implications of transfer pricing are being given heightened profile within corporate decision-making processes, with 28 per cent of respondents indicating that these issues are considered at boardroom level (Ernst and Young, 1997).

Because of the sensitive and confidential nature of these issues, there is little published research providing empirical evidence of the strategic geographies of intra-firm royalty payments.[1] Yet, there seems considerable likelihood that for large companies, the potential global mobility of intangible assets is an important factor in their profit strategies. These issues are particularly apposite for the food industry, because of the prominence of globally recognised brand names. In Australia, offshore royalty payment taxation deductions amounted to A$1.306 billion in 1993-94, and A$1.318 billion in 1994-95. The Australian food, beverages and tobacco industry made offshore royalty payment taxation deductions of A$117 million in both years (Australian Taxation Office, 1995, 1996).

The capacity for brand assets to drive corporate strategy is exemplified in the case of the Coca-Cola Company, holder of the world's most valuable brand name (estimated in 1996 to be worth US$43 billion; see Badenhausen, 1996: 29). The corporate strategy of the Coca-Cola Company is constructed tightly around the protection of its intangible brand assets. The Coca-Cola Company does not actually manufacture Coca-Cola, but leaves this function to bottling companies tied by contract. A result of these arrangements is the construction of a complex web of international cross-shareholdings and payments between and among bottlers and the Coca-Cola Company.

In summary, increased global mobility of capital facilitates the promotion of intangible assets as strategic and tradeable commodities within globalised processes of capital accumulation. For companies with powerful intangible assets - such as Coca-Cola, Philip Morris or Kellogg - these processes are potent. With these points in mind, attention now turns to the operations of Nestlé.

Nestlé's global branding strategy

Nestlé is the world's largest food company and owns many of the world's most prominent food brands. According to the United Nations Conference on Trade and Development, Nestlé is the world's most 'internationalised' transnational corporation, with 87 per cent of its assets, 98 per cent of its sales, and 97 per cent of its workers outside Switzerland, its home country (*The Economist*, 1997b). Between 1992 and 1996, Nestlé's global sales grew 11 per cent, while net profit grew 26 per cent (Nestlé S.A., 1997: 6). This strong financial performance has rested on the company's ability to generate internal growth for its core products, supplemented through acquisitions and joint ventures. These have included a joint venture in the breakfast cereal industry with the US transnational food company General Mills, and a string of acquisitions in the mineral water industry (notably Perrier) that have made Nestlé the world's largest mineral water company.

Notwithstanding these developments in the 1990s, Nestlé's corporate strength remains in its traditional pursuits of beverages, milk products and confectionery. Together, these activities accounted for 69.5 per cent of Nestlé's global sales in 1996. Within these core activities, the beverage sector is the primary profit engine for the company. In 1996, Nestlé global beverage operations contributed 14.9 per cent of Nestlé sales, but 40.1 per cent of Nestlé trading profit (Nestlé S.A., 1997: 8). This profitability has been built on the company's ownership and exploitation of brand names. Nestlé invented soluble coffee in 1938 and owns the world's most popular coffee brand, Nescafé. In Australia, Nestlé controls over 70 per cent of instant coffee sales via its Nescafé and International Roast brands (Australia, Prices Surveillance Authority, 1994:

27). Nestlé's market dominance provides the company with considerable scope to build profit margins in instant coffee that are higher than food industry norms (Australia, Prices Surveillance Authority, 1994: 53-54).

The development of global brands such as Nescafé (and others such as Milo, Maggi, Minties, Perrier, Carnation and Friskies) is, however, just one arm of Nestlé's branding strategy. Nestlé complements its promotion of a few globally recognised brand names with an intense attention to the development of localised brands that attempt to fill specialised regional and product niches. The company owns over 8,000 brands worldwide, but only 750 of these are registered in more than one country, and just 80 are registered in ten or more (Rapoport, 1994: 194). This tendency for the vast majority of Nestlé's brands to be registered in just one country means that Nestlé's 'supermarket shelf' presence differs greatly between countries. According to a supermarket survey conducted in 67 countries between 1993 and 1995, only 20 Nestlé brands

Table 2.1   The distribution of 701 Nestlé brand names in 67 countries

	Nestlé brand names present in:		
	1-3 countries	4-33 countries	over 34 countries
Confectionary	97	47	4 (Crunch, Kit Kat, Nestlé, Smarties)
Cereal	39	15	-
Coffee	65	28	3 (Nescafé, Decaf, Classic)
Infant Formula	22	19	2 (Cerelac, Nan)
Milk	47	13	4 (Carnation, Coffee Mate, Nestlé, Nido)
Miscellaneous dry goods	30	14	1 (Libby)
Refrigerated/ frozen goods	85	17	-
Petfoods	21	7	1 (Friskies)
Toiletries	72	43	5 (Elnett, Elseve, Free Style, Plenitude, Studio Line)

Source: Boze and Patten (1995: 27-32)

were identified as being used in 34 or more countries (Boze and Patten, 1995) (Table 2.1).

Evidently, the prominent role of local brands within Nestlé's brand armoury is part of a deliberate intent to attain both a global and local presence in the minds of consumers. This strategy is described candidly in a 1996 interview with the CEO of Nestlé Brazil, who identifies the 'handling of corporate and strategic brands' as the primary issue for the company in the next ten to fifteen years (Parsons, 1996b).

This concept is also highlighted in an article published in the business magazine *Fortune*:

> After plodding along as a multinational purveyor of coffee, chocolate and milk, Nestlé is blossoming into a brand franchise machine that lives mostly on local brands. The company prefers brands to be local and people regional - only technology goes global...In the developing world, it grows by manipulating ingredients or processing technology for local conditions and then slaps on the appropriate brand name (Rapoport, 1994: 137).

However, this pursuit of local brand niches is linked to broader goals of strengthening the Nestlé corporate brand. Often, product packaging or advertising retains some vestige of Nestlé's corporate identity, either through explicit use of the word 'Nestlé', variations of this word (such as 'Nestea', or 'NesQuik'), or via use of the 'bird's nest' corporate logo. The use of a unifying corporate identity is relaxed only in those cases where Nestlé products have strong brand identities in their own right, or where an association with Nestlé could detract from the product's appeal for particular market segments (the example of Perrier being an obvious case in mind). This cross-promotion of the Nestlé corporate brand differs from the approach taken by some other transnational food companies. For example, Unilever tends not to highlight the corporate parentage of its subsidiaries (Boze and Patten, 1996: 32).

The strong unifying presence of the Nestlé corporate brand within the company's range of food products is mirrored in the company's organisational management of its brands. Strategic brand positioning is orchestrated from the company's head office at Vevey, Switzerland (Parsons, 1996a). In general, Nestlé brands are owned by the group parent, Societe des Produits Nestlé SA, Switzerland, and licensed back to Nestlé subsidiaries (Perrier, 1997: 108). The centralisation of Nestlé's brand ownership structure may be linked to global finance and tax planning strategies. One commentator has noted that, following its 1988 takeover of the UK confectionery company Rowntree PLC,

Nestlé relocated the Rowntree brands to Switzerland 'for more favourable tax treatment' (Ball, 1994).

Nestlé Australia and Peters ice cream

Nestlé Australia is a wholly owned subsidiary of Nestlé's global parent, Societe des Produits Nestlé S.A. In August 1995, Nestlé Australia completed a A$570 million deal to acquire the assets of the Australian ice cream manufacturer Peters. Nestlé purchased Peters from the Australian transnational corporation Pacific Dunlop which, a few months earlier, had announced its intention to sell its food businesses (Pritchard, 1995). The acquisition of Peters marked the entry of Nestlé into the Australian ice cream industry. This acquisition also fulfilled Nestlé's desires to expand its ice cream operations in the Asia Pacific. In 1992, Nestlé made an entry to the Chinese ice cream market via a US$167 million (A$239 million) acquisition of ice cream businesses in Southern China (Clifford, 1992).

At the time of the Nestlé acquisition, the Peters ice cream business was a large, domestically focused ice cream manufacturer. The company had extensive reach into each of the three distinct segments of the ice cream market. In branded product 'stick' sales, it vied with the Unilever-owned Streets ice cream company for market share. For many years, this component of the industry has been a mature duopoly. Streets and Peters have attempted to mirror the other's product range, so that Streets' 'Cornetto' competes against Peters' 'Drumstick'; Streets' 'Magnum' competes against Peters' 'Heaven on a Stick', and Streets' 'Splice' competes against Peters' 'Mivvi'. (However, in the 1996-97 Australian summer, Cadbury-Schweppes entered this market, potentially destabilising the market shares of Streets and Peters).

In 'tub' sales, Peters vies with Unilever and a number of smaller manufacturers (including Kitchens of Sara Lee and Norco) for supermarket sales. In the 1990s, this market segment has seen the entry of a number of 'up-market' brandings that has sharpened the distinction between premium-priced product and the generic product. In food service, Peters held a number of lucrative supply contracts with companies such as McDonald's.

Nestlé's purchase of the Peter's business kindled media interest in the company's aspirations for its new acquisition. In particular, media scrutiny focused on Nestlé's capacity and interest in making export sales of Australian-made Peters ice cream. The former owner of Peters, Pacific Dunlop, had made extravagant claims of the company's export potential, though its own results were disappointing. The notion of the Australian food industry as a 'supermarket to Asia' is a dominant theme that cuts across much popular, bureaucratic and academic critique of Australian food industry restructuring (Pritchard, forthcoming). Nestlé's immediate actions after taking control of Peters were to close an ageing Peters' factory in Brisbane, and to recapitalise the company's core production facility in suburban Melbourne. It was announced that Nestlé would spend A$67 million in capital investment at the Melbourne plant between 1996-2001. At the time of writing, it is perhaps premature to judge Nestlé's motivations for this capital investment. However, it is clear that for the foreseeable future, the Peters ice cream business will earn the vast majority of its revenue from sales to Australian customers. In this respect, the management of Peters' ice cream brands becomes a central issue within Nestlé's profit strategy for the company.

Through its 1995 acquisition of Peters, Nestlé gained ownership of a valuable stable of product-specific ice cream brand names (Table 2.2). Most of these were developed in Australia for the local marketplace (the main exception being 'Drumstick', which was developed in the United States and

Table 2.2   Six most prominent Peters ice cream brands

Trademark (trademark number)	Date first registered	Date trademark assigned to Societe des Produits Nestlé SA
Icy-Pole (59348)	28 January 1932	3 December 1996
Billabong (295821)	8 April 1976	3 December 1996
Frosty Fruits (571750)	4 February 1992	3 December 1996
Choc Wedge (482024)	18 November 1954	3 December 1996
Drumstick (118375)	14 May 1954	13 May 1996
Monaco Bar	not registered	-

Source: Australian Intellectual Property Office (1997)

manufactured by Peters under licence). The highest selling Peters ice cream, Icy-Pole, was first registered in 1932. In addition, the corporate brand 'Peters' had wide recognition and value. In general, 'stick' ice creams are co-branded (that is, a company name complements a product brand), which allows various products to be advertised under an umbrella corporate banner. As a result, a single corporate name (Peters) can be used for signage on shopfronts, delivery vans, and freezer cabinets in corner stores. Traditionally, Peters has made extensive use of its corporate brand, through the logo: 'When it's this good it's Peters'.

Nestlé's acquisition of Peters raised some tangible questions for the company's strategic use of these brand names and, in particular, the use of the Peters corporate brand. Evidently, the Peters corporate name could be jettisoned, with its place taken by the Nestlé name and 'bird's nest' logo. This course of action would optimise exposure of the Nestlé name, enhancing its brand value for the company generally. However, to jettison the Peters brand would necessitate the net asset value of this brand being written-off, and potentially a loss of the consumer goodwill associated with the Peters' brand name. Alternatively, Peters could be retained without mention of Nestlé. This is the strategy adopted by Unilever, which uses the Streets brand as a corporate identifier. However, this course of action would appear less attractive for Nestlé than it does for Unilever. Nestlé has a strong consumer association with milk products, which extends readily to ice creams.

Nestlé has adopted a compromise strategy between these alternative positions. In February 1996, Nestlé registered for trademark purposes a logo that amalgamated the Nestlé and Peters corporate names. This logo incorporates the Peters' traditional green corporate identifier within Nestlé's traditional blue corporate identifier. Symbolically, the 'Peters' is situated within an encompassing 'Nestlé' oval, evoking a sense that Peters has a subordinate status within Nestlé's corporate ownership. This logo was launched for the 1996-97 summer season.

This brand strategy has clear parallels to that used by Nestlé in other markets. Nestlé has retained the large number of Peters' brands that have an Australian cultural resonance and which fill local market niches (for example, 'Billabong'), but has linked these to an overarching corporate parentage. This strategy aims to exploit the high visibility of ice cream branding, particularly in corner shops. The need to store ice creams in freezer cabinets has led to the creation of specific techniques for ice cream promotion. Generally, ice cream companies provide corner shops with freezer cabinets free of charge, so long as agreed sales volumes are met. This practice helps tie particular shops to a single ice cream company.[2] However, because cabinets are marked with a corporate identifier, this strategy also provides a prominent vehicle for corporate promotion. Therefore, adding the Nestlé identifier to Peters' products vastly enhances the visibility of the Nestlé name in the community.

The future of these arrangements is uncertain. There is some evidence that Nestlé is planning to reduce its global number of brand names by progressively phasing out some local brands in favour of the Nestlé logo, as part of a wider agenda to promote multiple sourcing arrangements (International Union of Food Workers, 1996). Although stick ice creams may appear an unlikely candidate for international trade, increased international sourcing has occurred over recent years. This tendency is linked to the further expansion of global product brand names, such as 'Magnum'. During the 1997-98 Australian summer, Unilever imported 'Solero' ice cream from Denmark, and Nestlé/Peters imported 'Mivvi' from Spain. In Peters' case, it is certainly possible that increased use of global product brands and global sourcing may encourage an eventual demise of the Peters corporate identity.

A second string of Nestlé's brand strategy for Peters ice cream has been to restructure the intra-corporate chain of ownership for the Peters brands. In December 1996, approximately 15 months after taking control of Peters, Nestlé reassigned the ownership of Peters' registered trademarks to the group's parent company Societe des Produits Nestlé SA (Table 2.2). Previously, Nestlé's Australian subsidiaries, Petersville and Peters Foods, owned these trademarks. There was no legal necessity for Nestlé to shift these trademarks to its group parent company: this represents Nestlé's global brand management strategy in action.

There are significant potential implications of assigning the Peters' trademarks to Societe des Produits Nestlé SA, in terms of intra-firm financial transactions. Figure 2.1 illustrates the chain of trademark ownership that Nestlé inherited upon acquiring Peters ice cream. At this time, the manufacturer of ice cream products (Petersville and Peters Foods) also owned the trademarks for those products. As a result, there was no potential flow of intra-firm royalty payments. Figure 2.2 displays the situation subsequent to December 1996. The assignment of trademarks to the Swiss-incorporated Societe des Produits Nestlé SA separates the ownership of trademarks from the ownership of production facilities and, as a result, opens a potential avenue for international intra-firm royalty payments. These payments may provide an alternative vehicle to dividends, as a means of repatriating monies from Nestlé Australia

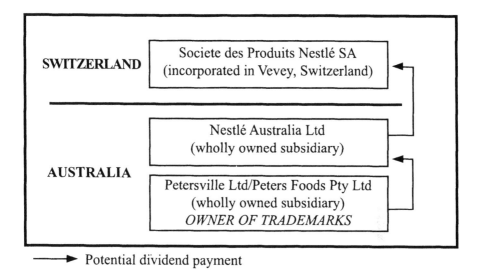

Figure 2.1   Nestlé/Peters potential ice-cream profit repatriation/
trademark royalty flows, August 1995-December 1996

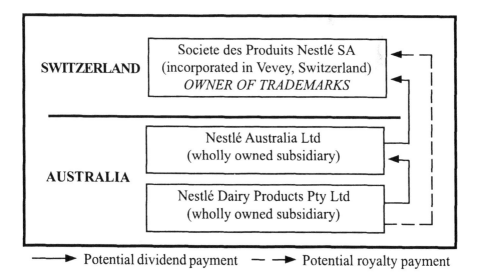

Figure 2.2   Nestlé/Peters potential ice-cream profit repatriation/
trademark royalty flows, after December 1996

to its Swiss parent company.

From published corporate records, it is unclear whether Societe des Produits Nestlé SA makes a royalty charge for the use in Australia of its Peters' trademarks. However for Nestlé Australia's operations as a whole (that is, confectionery, beverages, dairy products, pet food and ice cream), substantial international royalty payments are made to Nestlé group companies. According to Australian Securities Commission records, in 1996 Nestlé Australia made A$35.137 million of royalty payments to Societe des Produits Nestlé SA, and a further A$10.729 million of royalty payments to other Nestlé group companies. In total, these payments increased by A$5.5 million from their level the previous year, a period corresponding with the assignment of Peters' trademarks to Nestlé Australia's Swiss parent. Nestlé Australia's annual return indicates that these payments are made at an arms length basis (Nestlé Australia, 1997: 29).

Conclusion

This paper has provided a preliminary sketch of some of the issues relating to the global economic geography of brand names and intangible assets. Intangible assets have received scant attention in much of the published research into the globalisation of the agri-food system. Through the mobility of finance capital, new 'possible geographies' are opened to transnational corporations with respect to the ways brand names are owned and managed. A case study of Nestlé, the world's largest food company, reveals a complex and highly sophisticated global brand strategy. This involves an articulation of local and global brands nested within a brand ownership structure centred on the company's home country of Switzerland. Elements of this global brand strategy have been in evidence following Nestlé's 1995 acquisition of the Australian ice cream producer, Peters. A result is that brands with local consumer meanings, such as 'Billabong', are now held in Switzerland.

Such issues pose pressing theoretical and empirical questions relating to global governance and social regulation of the world food order. It is clear that globalisation processes in the agri-food system reach beyond issues of product sourcing, sales and production. The agri-food research agenda must confront issues concerning intangible assets such as brand names, if processes of globalisation are to be theorised comprehensively and patterns of global agri-food restructuring documented with accuracy.

Notes

1   Specific literatures in fields such as taxation accounting and corporate law contain considerable reference to these issues. However, these literatures are oriented mainly for practitioners in these fields and, in most cases, merely provide a commentary on the specific details of particular cases.

2   Under the Trade Practices Act, ice cream companies cannot insist that shop owners use freezer cabinets for their products only. However, by linking freezer provision to agreed sales volumes, shop owners are provided with disincentives to use the freezers for competitors' products.

References

Australia, Commissioner of Taxation (1997a), 'Definition of Royalties', *Income Taxation Ruling IT 2660*, 31 October.

Australia, Commissioner of Taxation (1997b), 'Ruling, Obligations of Royalty Payers', *Income Taxation Ruling IT 2669*, 31 October.

Australia, Prices Surveillance Authority (1994), *Inquiry into the Instant Coffee Declaration*, Australian Government Publishing Service: Canberra.

Australian Intellectual Property Office (1997), unpublished trademark ownership identification documents obtained by the author.

Australian Taxation Office (various years), *Taxation Statistics*, Australian Government Publishing Service: Canberra.

Badenhausen, K. (1996), 'There's Nothing Wrong with Being Number Two', *Financial World*, Vol. 165, No. 6, p. 29.

Ball, M. (1994), 'The Brand is Dead, Long Live the Brand', *Corporate Finance*, No. 119, p. 40.

Boze, B. and Patten, C. (1995), 'The Future of Consumer Branding as Seen From the Picture Today', *Journal of Consumer Marketing*, Vol. 12, No. 4, pp. 20-41.

Bryan, R. (1995), *The Chase Across the Globe*, Westview Press: Boulder.

Clifford, M. (1992), 'Ice Cream, Comrade?', *Far Eastern Economic Review*, 9 July, p. 67.

Connell, J. and Pritchard, W. (1990), 'Tax Havens and Global Capitalism: Vanuatu and the Australian Connection', *Australian Geographical Studies*, Vol. 28, No. 1, pp. 38-50.

Corbridge, S. and Thrift, N. (1994), 'Money, Power and Space: Introduction and Overview', in Corbridge, S., Martin, R. and Thrift, N. (eds), *Money, Power and Space*, Basil Blackwell: Oxford, pp. 1-26.

Donaghu, M. and Barrf, R. (1990), 'Nike Just Did It: International Subcontracting and Flexibility in Athletic Footwear Production', *Regional Studies*, Vol. 24, No. 6, pp. 537-552.

The Economist (1997a), 'Outsourcing: Separate and Lift', 20 September, p. 81.

The Economist (1997b), 'Multinationals', 27 September, p. 119.

Ernst and Young (1997), *Transfer Pricing 1997 Global Survey*, Ernst and Young International: Melbourne.

Hirst, P. and Thompson, G. (1996), *Globalization in Question: The International Economy and the Possibilities of Governance*, Polity Press: Cambridge.

Hughes, A. (1996), 'Retail Restructuring and the Strategic Significance of Food Retailers' Own-Labels: a UK-USA Comparison', *Environment and Planning A*, Vol. 28, pp. 2201-2226.

Interbrand (1997), at <http://www.interbrand.inter.net/Who/Docs/phil.html>, 27 October.

International Union of Food Workers (1996), *Nestlé Bulletin*, No. 43: Geneve/Pt-Lancy, Switzerland.

Leyshon, A. (1996), 'Dissolving Difference? Money, Disembedding and the Creation of Global Financial Space', in Daniels, P. and Lever, W. (eds), *The Global Economy in Transition*, Addison Wesley Longman: Harlow, pp. 62-82.

McDonald's Corporation (1996), *Annual Report*, McDonald's Corporation: New York.

Napier, C. (1994), 'Brand Accounting in the United Kingdom' in Jones, G. and Morgan, N. (eds), *Adding Value: Brands and Marketing in Food and Drink*, Routledge: London, pp. 76-102.

Nestlé Australia (1997), Annual Return to the Australian Securities Commission, ASC document 00001131G.

Nestlé SA (1997), *Annual Report*, Societe des Produits Nestlé: Vevey, Switzerland.

Parsons, A. (1996a), 'An Interview with Joe Weller, Chairman and CEO, Nestlé USA', *McKinsey Quarterly*, No. 2, pp. 14-17.

Parsons, A. (1996b), 'An Interview with Roland Meyes, President and CEO, Nestlé Brazil', *McKinsey Quarterly*, No. 2, pp. 22-25.

Perrier, R. (1997), 'Brand Licensing', in Hart, S. and Murphy, J. (eds), *Brands: The New Wealth Creators*, Macmillan: London.

Pritchard, W. (1995), 'Foreign Ownership in Australian Food Processing: the 1995 Sale of the Pacific Dunlop Food Division', *Journal of Australian Political Economy*, No. 36, 26-47.

Pritchard, W. (forthcoming), 'Australia as the 'Supermarket to Asia'? Governments, Territory and Political Economy in the Australian Agri-food System', *Rural Sociology*.

Rapoport, C. (1994), 'Nestlé's Brand Building Machine', *Fortune*, Vol. 130, No. 6, pp. 147-56.

Roberts, S. (1994), 'Fictitious Capital, Fictitious Spaces: the Geography of Offshore Financial Flows', in Corbridge, S., Thrift, N., and Martin, R. (eds), *Money, Power and Space*, Blackwell: Oxford, pp. 91-115.

Sara Lee Corporation (1996), *Annual Report*, Sara Lee Corporation: Chicago.

Sparks, L. (1997), 'From Coca-colonization to Copy-cotting: the Cott Corporation and Retailer Brand Soft Drinks in the UK and the US', *Agribusiness*, Vol. 13, No. 2, pp. 153-167.

Tickell, A. and Peck, J. (1992), 'Accumulation, Regulation and the Geographies of Post-Fordism: Missing Links in the Regulationist Research', *Progress in Human Geography*, Vol. 16, No. 2, pp. 190-218.

Whatmore, S. and Thorne, L. (1997), 'Nourishing Networks: Alternative Geographies of Food', in Goodman, D. and Watts, M. (eds), *Globalizing Food*, Routledge: London, pp. 287-304.

3 An apple a day: Renegotiating concepts, revisiting context in New Zealand's pipfruit industry

MEGAN McKENNA, MICHAEL ROCHE AND RICHARD LE HERON

Introduction

As consumers of food, most people are only marginally aware of how food is produced and have little knowledge of the social, economic, political and environmental issues surrounding conditions of production.[1] Consumer purchasing decisions for fresh fruit and vegetables for example, are usually based on appearance, perceived 'freshness' and quality, and price. The construction of food 'quality' and 'image' reflects complex, dynamic processes drawing on (among other things) discourses of 'sustainability' and 'restructuring'. That these discourses are differentially interpreted and contested at various organisational/spatial scales is central to our investigation of global-local links and regime restructuring in New Zealand's apple industry.

This chapter reflects on two main themes. The first theme highlights the need to revisit the content of key theoretical categories and conceptual challenges associated with food regimes research. Typically, geographic literature on food regimes and food complexes has remained silent about how restructuring processes and regulatory mechanisms shape place-bound responses to sustainability discourse and debates. Stemming from these issues, the second theme considers how the multi-scaled functions of industry-specific regimes are 'placed' in terms of locality-based and individually-practiced production strategies. That these production strategies are sometimes inherently conflictual between growers' economic, social and environmental ideologies (and constraints) and the production and organisational guidelines pursued by the statutory authority controlling the pipfruit industry, emphasizes the

41

importance of locality in understanding how food regimes are defined and contested.

Discussion of the main themes follows a brief description of central theoretical concepts and an introduction to our research focus on New Zealand's apple industry. The intent here is to review key aspects of the conceptual architecture underlying geographic notions of processes and relationships in the world food economy. In this chapter, the emphasis is placed on 'locality' as the focal point for negotiating the spatially complex, socially constituted and mutually recursive nature of global-local linkages that shape aspects of production-consumption relations in the global fresh fruit and vegetable (FFV) complex.

New Zealand (sustainable) apples and the political economy of food production/consumption

Apples have a long cultivation history, having been grown as a domestic crop for nearly 3,000 years. The versatility of apples as a crop and their consumer appeal as a fruit is reflected in the geographical spread of the leader producer regions (Tables 3.1 and 3.2). China is the leading producer, having been equal with the US as recently as 1990. France, the third ranked producer in 1997 (at half US levels) and the remaining top ten are spread throughout Western and Eastern Europe, the Middle East, South Asia and Latin America.

Only some of the leading producers export sufficient quantities to rank in the top 10 exporting nations, notably France, Italy and Argentina. The proportion of the crop exported by these top ten countries ranges from 20 to 30 percent for Italy and France but as low as 14 percent for Argentina. Some major producing areas have minimal exports in a percentage and absolute sense and are still comparatively self sufficient in terms of domestic apple markets.

Southern hemisphere apple producers occupy four of the top ten positions in the list of leading exporters. More significant perhaps is the export orientation of their those countries' apple industries. Thus, while New Zealand contributes only about one per cent of world apple production and two to three per cent of world apple exports, over 50 percent of its production is exported. Chile exports 50 percent of its production, while with South Africa the degree of export specialisation is less marked, but still significant at 33 percent.

The previous discussion of world apple production and export patterns

Table 3.1   Leading apple producing countries, 1996

	Production Tonnes ('000)	Exports Tonnes ('000)	Percent exported
China	17,056	165	<1
US	4,714	615	13
France	2458	827	34
Turkey	2,200	56	2
Italy	2,071	448	21
Germany	2,060	48	2
Poland	1,952	95	4
Iran	1,925	190	9
Russia	1,800	1	<1
Argentina	1,300	188	14

Source: FAO (1996a).

Table 3.2   Leading apple exporting countries, 1996

	Exports Tonnes ('000)	Production Tonnes ('000)	Percent exported
France	828	2,458	34
US	615	4,714	13
Italy	448	2,071	21
Chile	442	880	50
Netherlands	377	490	77
Belgium*	322	299	108
New Zealand	296	567	52
South Africa	209	631	33
Iran	190	1,925	9
Aregentina	188	1,300	14

Source: FAO (1996b).
* Includes Luxemburg and re-exports

has tended to present apples as an undifferentiated commodity, but in fact, the production of apple varieties is quite varied. To some extent this is a function of biophysical and historical episodes in the development of new varieties. New Zealand growers have however, tried to position themselves as leading exporters of 'newer' innovative varieties (notably Braeburn and Royal Gala as shown in Table 3.3), and have restructured the national export crop away from traditional varieties (Red Delicious, Golden Delicious, Granny Smith) still produced in large volumes by other leading exporters. The combination of producing high quality, alternative varieties and being a southern hemisphere producer has helped New Zealand develop a preferred supplier status with large, northern hemisphere markets in the EU, UK and US. With the advent of improved cool store technology in key markets (particularly the EU), the advantage of being a southern hemisphere supplier has eroded somewhat. In

Table 3.3   **Top apple varieties as a percentage of production and exports, US, EU and New Zealand, 1994-96**

	US (production)	EU (production)	New Zealand (exports)
Red Delicious	39	12	8
Golden Delicious	14	39	-*
Granny Smith	6	5	6
Rome	5	-	-
Fuji	5	-	9
Jonagold	-	10	-
Elstar	-	4	-
Gala	3	3	3
Cox's Orange Pippin	-	2.7	6
Royal Gala	-	-	22
Braeburn	-	-	40

Source: Anonymous (1995a; 1995b; 1996), NZAPMB (1996).

* Top varieties only are represented; a dash does not indicate a variey is not produced.

addition, given the current oversupply of most fruits in world markets, including apples, it is increasingly important to ENZA (the marketing arm of the New Zealand Apple and Pear Marketing Board) to differentiate New Zealand apple varieties and supply parameters from its competitors. The key business drivers of ENZA's export programme for 1998 includes a shift from apples being 'fit for export' towards being 'fit for market' (NZAPMB, 1997) - an approach which revolves around a new and rigorous export specification profile (ESP) determining fruit colour banding, size and fruit pressures.

Producing fruit to environmentally acceptable standards is an increasingly important market access issue. The integrated fruit production (IFP)[2] approach to growing pipfruit is also clearly linked to ENZA's marketing imperatives as well as evolving in the context of a national legislative/ideological shift towards sustainable management practices in New Zealand agriculture. The year before the IFP pilot programme was initiated (1996-97), involving about 80 growers in three different producing regions (Hawkes Bay, Nelson and Central Otago), a consulting report noted:

> Many European supermarket chains have advised ENZA that unless they can supply IFP grown product as soon as 1997 they will not be considered preferential suppliers and risk the possibility of not being able to supply their market. There is no back tracking on this (Wilson, 1996:1).

The Ministry of Agriculture (MAF) outlined its position on sustainable agriculture in a 1993 policy paper, which drew on the Resource Management Act (1991) and the Biosecurity Act (1993). In general, MAF's approach aims to ensure profitable conditions for primary producers, through better land management, which may contribute to a 'narrower' adoption of sustainability principles than some other ministries (Blunden *et al.*, 1996). The focus on 'management' issues raises several questions about defining and implementing sustainable farm practices, including: the need to recognise that 'local' sustainable management options may have implications for environmental change that extend beyond individual and community control (Blunden *et al.*, 1996); the need for institutional and bureaucratic support of sustainability principles and initiatives; and the need to anticipate and/or accommodate ways in which sustainability concepts are inherently conflictual (from spatial, regulatory, cultural and individual points of view) and may be appropriated by different sets of priorities (social, economic, environmental) and values.

Set against a backdrop of problematising 'sustainability' and the differential insertion of food producing regions/industries into the global FFV complex,

ENZA and the national grower cooperative is faced with the challenge of shaping industry-specific regulatory guidelines towards the construction of apples as a food commodity that is 'fresh', 'sustainable' and attractive to consumers. From a theoretical point of view, some of these relationships are captured in food regimes research in agricultural political economy (Friedman 1993; Friedman and McMichael, 1989). 'Food regimes' refer to the multi-scaled structuring of food production and consumption governed by axes and sets of rules (Le Heron, 1993). In the case of the FFV complex, emerging sets of rules and socio-structural characteristics revolve around constructions of 'fresh', 'sustainable' and the consumption of a 'quality' product. A useful contribution of food regimes research has been to open up key debates about how local (sub-national and regional) farming practices, social organisations and 'rural lifestyles' are being restructured and realigned throughout the world.

Although difficult to define, the transition to a third, consumption-driven food regime in the world food economy appears to be characterised by relations of 'globalisation' and sustainability (Arce and Marsden, 1993; Jarosz, 1996; Le Heron and Roche, 1995a; 1995b; Marsden and Arce, 1995; Marsden *et al.*, 1996; Ward, 1993; Whatmore, 1995), and the stretching over space of power, valorisation and productive relations (Jarosz, 1996; Massey, 1994). New conceptual spaces in food regimes research have been opened up through questions about regional specificity in the value construction of food, farm practices and industry regulatory mechanisms (Goodman and Watts, 1994; Marsden *et al.*, 1992; Marsden and Arce, 1995; McMichael, 1996; Moran *et al.*, 1996).

Our research suggests that while notions of 'sustainability' historically coincide with the reconstitution of the world food economy towards 'fresh', 'safe' and 'organic', the contemporary lived experiences of New Zealand orchardists highlight how economic ideology has appropriated and interpreted 'sustainable' discourse. Further, these lived experiences are locality-specific. Local and industry adaptation of sustainability principles are constrained by the product (in this case pipfruit) and national/international circumstances (political, economic and social), which to a large degree have converted sustainability debates into a discussion creating economic advantage for New Zealand pipfruit in global markets. A critical renegotiation of 'macro-scale' processes (for example those structuring the world food economy) may proceed on the basis that 'globalisation' is a historically-grounded (and shifting) conceptual framework, that needs to be unpacked to better understand how 'the local' shapes highly differentiated food production processes and consumption patterns/preferences.

Renegotiating the links between meta-theoretical constructs and place-specific practices and processes in our research was achieved through participatory-observation work and twenty intensive interviews with orchardists in three different growing regions (Hawkes Bay, Nelson and Central Otago). The interviews served two main purposes: firstly, to help expand our general knowledge of orchard-level operations in different regional contexts; and secondly, to gain insight into 'locality' as an empirical and conceptual focal point for understanding how 'globalisation', sustainable discourse and the dynamics of the 'world food economy' are given meaning through lived experiences. Interviews took place on the orchards and were combined with orchard walks. Discussions were semi-structured, with the interviewer introducing key research themes and allowing the orchardists to interpret meanings, relevance and specific examples according to their experiences. Here, we hoped to gain insight into the simultaneous constitution of global-local processes as expressed through production issues, industry structure and the perceptions of fruit growers.

Table 3.4 outlines some of the key links in our project of renegotiating (theoretical) concepts and categories and 'placing' processes. The table reflects our approach to connecting analytical concepts (reading the columns left to right) with meta-theory and processes which are 'placed' in the context of New Zealand's apple industry. This approach allows us to explore the importance of 'restructuring' regimes and locality-based difference in examining: how food producing complexes and regimes represent temporarily fixed structures/meanings shaped by local agency and regulatory mechanisms; and how macro-scale theory and process have meaning in terms of locality-based constraints and opportunities for food producers. Underlying both of these points is the necessity to 'people' food regimes research and reiterate the fact that food producing systems are socially constituted.

Increasingly, concepts of sustainability and food quality are entering into consumer choices and agricultural political economy theory. The complex and often vague nature of sustainability concerns were reflected in growers' wide-ranging interpretations of 'sustainable agriculture' and reasons why they did (or did not) favour IFP production practices. Precisely because sustainability concepts are inherently conflictual (from spatial, regulatory, cultural and individual points of view), 'sustainable priorities' appear to differ significantly among growers, and to a certain extent, between orcharding regions. This highlights the utility of localities research in 'placing' concepts and processes, and in problematising the differential insertion of food producing regions/industries into the global FFV complex.

Table 3.4 Growing global, living local: Renegotiating categories, 'placing' processes

Concepts and Categories	Meta-theory versus macro-processes	'Placed' constructs and entities	Renegotiation and locality-based processes
Food	Fresh fruit and vegetables (FFV) global complex	New Zealand apples	• Variety mixes • Quality construction • Production issues, profitability
World food economy	Emerging regimes	New Zealand Apple and Pear Marketing Board (NZAPMB)	• Contesting the efficiency and ownership of the Board • Debating the value of 'cooperative'
Localities	• New Zealand • Global competitors	• Hawkes Bay • Nelson • Otago	• Regional competitiveness • Uneven development/power • Relative marginality
Sustainability	Combining economic social, environmental factors	• Resource Management Act • Biosecurity Act • MAF 'sustainable agriculture • Integrated fruit production programmes	• Fruit quality/market access • Profitability/costs • Farm practices • Grower culture

Restructuring regimes, contesting sustainability

This section briefly discusses selected discourses of sustainability by examining different locality experiences in the Nelson, Hawkes Bay and Central Otago orcharding regions. Each example highlights the multi-scaled complexity of local constraints and opportunities, industry restructuring, and orchard level interpretations of what is 'sustainable'. Drawing on grower's perceptions (which are part of a larger survey and in-depth interviews of more than 20 orchardists in three regions), Table 3.5 highlights the connections between place, people and process in interpreting contested meanings and discourses of sustainability. Although the table represents only a 'snapshot' of key processes and experiences within various orcharding localities, it attempts to highlight significant regionally-based concerns linking political restructuring within the national cooperative (NZAPMB), with economic viability concerns and the place of 'environmental debates' within New Zealand's export market focus.

Nelson - sustainable environments and/or sustainable markets?

The IFP pilot programme, initiated by ENZA in 1996-97, included 12 pipfruit orchardists from Nelson (from a total of approximately 475 growers), 55 from Hawkes Bay (from an approximate total of 900 growers) and 21 from Central Otago (from roughly 150 growers). The primary reason for Nelson's relatively low participation rates stems from the relationship between economic priorities and socio-environmental specificity.

To date, Nelson growers have the highest success rate in producing for the US market (stemming from a lower incidence of key pests), which is the second most important export destination for New Zealand pipfruit (accepting about 20 percent of the export crop). USDA requirements for fresh fruit stipulate zero tolerance levels for actionable pests, which means that growers producing for American markets follow conventional fruit production (CFP) practices that require regular applications of organo-phosphate sprays. Combined with 'zero tolerance' guidelines is the fact that USDA standards do not accept fully IFP-produced fruit.

Because of the strict US code of practice (which has been criticised by some growers as an invisible tariff barrier), and the importance of maintaining a preferred supplier status for that market, ENZA offers a NZ$1 premium per tray carton of fruit destined for America. The result is, that despite an overall industry (and international market) push towards fewer chemicals, Nelson,

Table 3.5 Restructuring regimes and locality-based discourses of 'sustainability': Linking people, place and process

Place	People	Process	Contested issues
Nelson	Gary: There are a lot more questions about the Board in Hawkes Bay...and there is more loyalty here. If the Board said 'go and jump off that hill over there', a lot of growers would. What I mean is, that there is a lot of loyalty, but not necessarily to IFP being pushed on us. We know that the market requirement is that we have to put fruit into the US, so Nelson growers do that... Claire: We took up IFP because we wanted to reduce the chemical input into the environment - fewer organo-phosphates. There is a financial incentive as well.	Sustainable environments and/or sustainable markets	• How is 'fresh' constructed? • Does sustainable really mean profitable? • Selling or protecting the environment?
Hawkes Bay	Roger: I was in the lift with [Associate Minister of Agriculture] Luxton yesterday, and I mean the single desk is gone. Finished. Government is going to get out of this business and a bloody good job too. Why should I, the farmer, put up with that crap from 49 years ago? I mean, 49 years ago Stalin was in power and there weren't any supermarkets!	Sustainability of what and for whom?	• Inherent conflicts in cooperative regime over global markets/ growers' duties • Grower 'cooperative culture' • Are cooperatives sustainable?

Place	People	Process	Contested issues
	Paul: I think the Board structure is brilliant. I support it wholeheartedly. Deregulation is driven by greed…When people do things as a group, they aren't driven by greed, they're actually driven by the safety and security and the practicality of marketing in that way.		
Central Otago	Keith: I had a gutful with ENZA, so after deregulation we went with the local market. The prices were pretty good becasue we were isolated…But, the problem with restructuring in the pipfruit industry now is that you can't just go down the road and sell your apples to the local supermarket, like before. There are a lot more chains now too. Supermarkets don't want to deal with 20 little suppliers when they could deal with one or two big ones. So, you pack your apples here, transport them to Christchurch, where they are bought and redistributed back to the store down the road. With prices like they are now, who is going to do that for much longer?	Sustaining the single desk and the conditions of marginality?	• Accommodating productive/economic marginality • Uneven distribution of political/social/ economic power

New Zealand's second largest growing region after Hawkes Bay, has been slow to adopt safer and more environmentally sustainable production practices. Added to these tensions is a historical competitive parochialism between Nelson and Hawkes Bay, and underlying suspicions among Nelson growers that IFP has been 'imposed' on them in a way which directly interferes with their profitability. 'Sustainable' discourse has, in this sense, strongly economic overtones.

As a way to address the regional and layered tensions of the IFP programme, and in direct response to demands by major European buyers of New Zealand fruit, ENZA has designed a new initiative offering a premium of NZ$0.25 on all IFP-produced export fruit in 1998. Negotiation with the United States has yielded and IFP-transitional programme (which also qualifies for the IFP premium) for 1998 and 1999. American markets will move towards acceptance of fully IFP-produced apples in the year 2000, by which time ENZA plans to have all growers (approximately 1,600 in number) practicing IFP techniques.

Hawkes Bay - sustainability of what and for whom?

There is a vocal minority calling for deregulation among growers - 95 percent of whom continue to support the notion of a national cooperative and single desk marketing structure. For the past four years, pipfruit growers have experienced declining profitability linked to increased international competition in traditionally 'New Zealand' apple varieties, an oversupply of fruit on world markets, increased production costs (particularly for packaging/ presentation) and relatively frequent and significant natural hazard events (notably hail storms and drought conditions). The current financial squeeze in the pipfruit sector has caused growers to look more carefully at what a national cooperative represents in terms of 'best chances' for the industry.

The interpretation of, and emphasis on, 'sustainability of what and for whom' shifts according to the desired changes of the industry's regime structure, which currently revolve around deregulation debates. Is the statutory desk viable and 'sustainable' in the context of intensive global competition and pressure to reduce producer subsidies? How 'sustainable' is New Zealand's apple industry without single desk marketing as an effective tool to promote growers' interests against larger, less geographically isolated competitors? How 'sustainable' are pipfruit-growing communities in the face of deregulation and industry downsizing? Explicitly structuring the range of opinions are personal and philosophical positions on how to best ensure (economic) sustainability within, or outside, the current regime.

Claims for immediate deregulation of the Board structure reflect two viewpoints: a decade of neo-liberal ideology by New Zealand's Labour and National governments; and the opinions of those growers most dissatisfied with the financial performance of ENZA in the last three selling seasons. 'Poor performance' is measured by final fruit value returns to growers, the apparent inability of marketing programmes to respond quickly to changing trade conditions, and the perceived inefficiency of ENZA's marketing costs (measured by percent of gross returns to the grower). For deregulationists, 'sustainability' means economic sustainability, something that should be contested and measured in a freer market environment.

Those expressing moderate and/or full support of the single desk system represent an industry vision equating the best chances of economic sustainability within the broader structures of the national cooperative. Here, speculative support remains for the Board, and the contested focus for change in the ways in which the current statutory Act is interpreted. For instance, there is room within the Act to grant individual growers private export licenses, a process which should (according to most growers) be made more accessible. Again, an economic rationale prevails in defining more sustainable industry options. Risk-taking growers with high quality, specialised products should have greater chances of reward than 'poor' growers, who represent the cooperative's lowest common denominator (defined by risk acceptance, willingness to innovate and product quality).

Of the vast majority of growers who support the single desk structure, few generalising statements can be made. Unlike deregulationists contesting the fundamental industry regime structure, Board supporters contest the level of ENZA's financial and operational accountability. Support for the statutory authority, although immediately 'economic', is based on a wider vision of the cooperative structure combining: perceptions of maximum personal and industry profitability, with explicit awareness of the place of orcharding as an economic and social activity within community structures.

Central Otago - sustaining the single desk and conditions of marginality?

Geographic isolation, small total production volumes, climatic conditions (dry with greater risk of frost than other regions) and less extensive, less accessible, packing and processing infrastructure make Central Otago growers relatively small players in New Zealand's pipfruit industry. In addition, most orchardists in Central Otago combine summerfruit (apricots, cherries, nectarines, peaches) with pipfruit production - pipfruit normally receiving secondary focus in terms of planted area. While there are concentrated pockets of pipfruit production at the southern end of the region, low industry profitability has stimulated increased discussion among many apple growers to diversify into summerfruit production by removing and replanting trees. These factors have helped shape a grower culture in Central Otago which is different from those of the larger, more centrally placed Nelson and Hawkes Bay.

The predominance of summerfruit in the region, which is not structured by a statutory authority, contributes to regionally defined experiences of being competitive, individualistic fruit growers. Lower tolerance levels of pipfruit's co-operative regime guidelines, rules and regulations is also fuelled by limited political influence on that regime resulting from Central Otago's historic marginality. In fact, some pipfruit growers focus on local market production, rather than export production, as a way to: opt out of the single desk structure (the local market was deregulated in 1993), and/or accommodate their more peripheral position within ENZA's export infrastructure.

Central Otago growers are faced with a complex set of political, economic and social conditions in evaluating the 'sustainability' of pipfruit production in the context of the current regime structure. Some growers have argued that the current industry organisation, and its stated trajectory to become more market-driven, entrench regional conditions of political-economic marginality. Further, doubts remain among Central Otago growers about whether, and how much, their overall industry position would improve (relative to other producing regions), even if the industry's financial picture looked more promising. This level of doubt about the longer-term prospects of Central Otago pipfruit orchardists has been translated into a clear political position endorsed by the two grower's associations (at Teviot and Ettrick), stating that their participation in, and recognition of, the current regime structure is on 'probationary' terms only for 1998. If the economic and political conditions governing the industry do not improve for Central Otago growers by the year end, growers have suggested they will opt out of the statutory guidelines and market their own fruit.

The present structure of the industry regime implies several levels of spatial and organisational marginality for Central Otago pipfruit growers. First, there is ENZA's restructured emphasis on being driven by global markets instead of domestic growers. Second, within the national co-operative infrastructure, no plans are underway by the Board to improve processing, packing and storage facilities. And third, local market supply networks are becoming increasingly centralised around Christchurch, making it more difficult and expensive for the local growers to sell fruit to local retail outlets (as the interviewee, Keith, notes in Table 3.5). In this case the conditions of economic, social and political sustainability for growers are complex, and in terms of day-to-day operations and financial viability, may depend upon where and how a consumer network for their product is defined and negotiated.

Conclusion

Until recently, the food regimes and global commodity chains literatures have been relatively silent on how place-specific social and political practices affect restructuring changes within industries, and how meanings of regime structures are emergent and temporarily fixed. The 'stable' patterns implicit in meta-theoretical thinking about food regimes needs to be contextualised in terms of political and social (which includes environmental) developments that reveal shifting power relationships in food imaging, valorisation and distribution. By extension, initiatives at particular sites in the chain (like the proposed changes to the ownership structure of the NZAPMB) can be contextualised so that the impacts and constraints of agency are better understood.

Focusing on locality-based interpretations of 'sustainability' highlights the conflictual and contested meanings of the concept in accordance with economic, political and social priorities. Renegotiation of the apple industry's regime structure and meanings of 'sustainability' within New Zealand's pipfruit sector highlight the importance of locality and regional difference in restructuring processes. The reassessment of food and sustainability means that the content and nature of economic processes are being problematised. Institutional structures and analytical categories are in flux, where old meanings no longer provided clear information and new meanings are still ambiguous in what they offer. Our research has shown that processes of contestation reveal place-specific attributes and tensions in peoples' interpretations of, and behaviours within, the spatially uneven dynamics of regime structures and functions.

Methodologically, the research explored at a simple level the construction of, and interpretation about, sustainability in three apple producing regions. It also investigated operational practices developed by growers against the backdrop of sustainability discourses. The chapter's evidence reveals that the nature of engagement in sustainability discourses is both problematic and negotiated. Analytically, discourses can be identified on the basis of agency connected with regulatory institutions embracing the apple industry (MAF, New Zealand Fruit Growers Federation, Ministry for Environment, New Zealand Apple and Pear Marketing Board) and economic agents (growers and post-harvest operators). ENZA's adoption of IFP was an overt marketing strategy to protect its market share in key regions such as the European Union. The communication of IFP may have been mishandled in the sense that various growing regions have different political and economic sensitivities to the introduction of a programme representing another industry standard. Growers in each producing region perceive sustainability issues in different ways, reflecting assumptions about market expectations that related to their export history and assessment of workable practices in the orchard. The growers perceptions of MAF's 'sustainability' agenda was minimal, and most growers had little idea of the connections between that agenda, the IFP programme and international marketing imperatives for safe foods.

The chapter shows that sustainability discourses are entangled in other discourses, contributing to regional tensions of a predominantly economic nature. In Nelson, debate over IFP has taken place in the context of maintaining the US code of practice. With overall industry profits declining, growers are attempting to maintain the prevailing (and lucrative) market arrangements. In Hawkes Bay, which has greater difficulty meeting the US code (primarily because of pest incidences), the sustainability discourse has been narrowed to the question of *who* represents the best option for sustainable participants in the industry. The debated alternative between a cooperative Board structure and a corporate structure translates into a very different mix of growers surviving to introduce and maintain sustainable practices. In Central Otago, growers' marginal status in the industry means the interest in IFP was in the immediate returns to the orchard, perhaps by differentiating their product as 'environmentally safe'. While IFP was viewed with some interest (or at worst, indifference), Central Otago growers were more preoccupied with their perceived tenuous and seemingly 'unsustainable' (political and economic) position within the national pipfruit industry.

By recognizing variation in growers' apprehension about sustainability ideas and practices, the discourses of sustainability can be seen as restructuring mechanisms in their own right. By resisting IFP the Nelson growers are specialising in one market, attempting to maintain returns but risking inflexibility in the face of market collapse. Hawkes Bay growers (along with the other regions) are confronting what grower features would be desirable under alternative regime structures. Growers in Central Otago face a starker dilemma - whether to participate in the industry's organised export sector at all.

Notes

1   We would like to acknowledge the support of the Massey University Post-Doctoral Fellowship programme and the Massey University Research Fund, for making this work possible. We are also grateful to all the growers who gave their time and shared insight into their lives, businesses and communities.
2   The pressure to adopt a set of sustainable management practices among pipfruit growers is reflected in the adoption of a more selective and less damaging spraying regime. Esentially, IFP outlines an integrated approach to pest and disease management; encourages monitoring to determine if pest and disease thresholds have been exceeded; and gives preference to non-chemical controls wherever possible. This is in contrast to the more 'traditional' blanket spraying programmes using organo-phosphate chemicals that operate on the non-selective basis of 'getting everything at once'. IFP also reflects an increased awareness and, in many cases, a long standing preference among growers, to follow more environmentally sensitive production practices.

References

Anonymous (1995a), 'Apples', *International Fruit World*, Vol. 53, No. 2, pp. 320-325.
Anonymous (1995b), 'Southern Hemisphere', *Deciduous Fruit Grower*, Vol. 43, No. 12, pp. 454-458.
Anonymous (1996), 'Apples at a Glance', *American Fruit Grower*, September, p. 7.
Arce, A. and Marsden, T. (1993), 'The Social Construction of International Food: A New Research Agenda', *Economic Geography*, Vol. 69, No. 3, pp. 293-311.
Bird, L. (1997) Ministry of Agriculture Policy Agent, Hastings, personal communications and interviews, December 1996-November 1997.
Blunden, G., Cocklin, C., Smith, W. and Moran, W. (1996), 'Sustainability: A View From the Paddock', *New Zealand Geographer*, Vol. 52, No. 2, pp. 24-34.
Buttel, F. (1996), 'Theoretical Issues in Global Agri-food Restructuring', in Burch, D., Rickson, R. and Lawrence, G. (eds), *Globalization and Agri-food Restructuring: Perspectives from the Australasia Region*, Avebury: Aldershot, pp. 17-44.
ENZA (1996a), *Annual Report*, New Zealand Apple and Pear Marketing Board: Wellington.
ENZA (1996b), *Strategic Plan*, New Zealand Apple and Pear Marketing Board: Wellington.
ENZA (1996c), *New Zealand Integrated Fruit Production - Pipfruit Manual*, New Zealand Apple and Pear Marketing Board: Wellington.
ENZA (1997a), *Road Show Presentation - Handouts*, Seminar to New Zealand growers in Hawkes Bay, Gisborne, Wairarapa, Nelson, Marlbourough, Otago, Christchurch, Auckland and Hamilton, July.

ENZA (1997b), *Proposed Ownership Structure*, presentation at the Pipfruit Growers Annual Conference, September 9-10: Wellington.

ENZA (1997c), *A Question of Ownership*, draft proposal document, November.

Friedmann, H. (1993), 'The Political Economy of Food: A Global Crisis', *New Left Review*, No. 197, pp. 29-57.

Friedmann, H. and McMichael, P. (1989), 'Agriculture and the State System: The Rise and Decline of National Agricultures, 1870 to the Present', *Sociologia Ruralis*, Vol. 29, pp. 93-117.

Food and Agriculture Organisation (FAO) (1996a), *Production Year Book*, FAO: Rome.

Food and Agriculture Organisation (FAO) (1996b), *Trade Year Book*, FAO: Rome.

Goodman, D. and Watts, M. (1994), 'Reconfiguring the Rural or Fording the Divide? Capitalist Restructuring and the Global Agro-Food System', *The Journal of Peasant Studies*, Vol. 22, No. 1, pp. 1-49.

Grosvenor, S., Le Heron, R. and Roche, M. (1995), 'Sustainability, Corporate Growers, Regionalisation and Pacific-Asia Links in the Tasmanian and Hawkes Bay Apple Industries', *Australian Geographer*, Vol. 26, No. 2, pp. 163-172.

Jarosz, L. (1996), 'Working in the Global Food System: A Focus for International Comparative Analysis', *Progress in Human Geography*, Vol. 20, No. 1, pp. 41-56.

Lawrence, G. (1996), 'Contemporary Agri-food Restructuring in Australia and New Zealand', in Burch, D., Rickson, R. and Lawrence, G. (eds), *Globalization and Agri-Food Restructuring: Perspectives from the Australasia Region*, Avebury: Aldershot, pp. 45-72.

Le Heron, R. and Roche, M. (1996a), 'Eco-commodity Systems: Historical Geographies of Context, Articulation and Embeddedness Under Capitalism', in Burch, D., Rickson, R. and Lawrence, G. (eds), *Globalization and Agri-Food Restructuring: Perspectives from the Australasia Region*, Avebury: Aldershot, pp. 73-90.

Le Heron R. and Roche, M. (1996b), 'Globalisation, Sustainability, and Apple Orcharding, Hawkes Bay, New Zealand', *Economic Geography*, Vol. 72, No. 4, pp. 416-432.

Le Heron, R. and Roche, M. (1995), 'A Fresh Place in Food's Space', *Area*, Vol. 27, pp. 22-32.

Massey, D. (1994), *Space, Place and Gender*, University of Minnesota Press: Minneapolis.

Massey University (various years), *Geography Department Grower Survey*, December 1996 - February 1998.

Marsden, T. and Arce, A. (1995), 'Constructing Quality: Emerging Food Networks in the Rural Transition', *Environment and Planning A*, Vol. 27, No. 8, pp. 1261-1279.

McKenna, M., Roche, M. and Le Heron, R. (1997a), 'Food Regimes Research and Labour Processes In Flux', *Proceedings of the Second Annual Meeting of the Institute of Australian Geographers and New Zealand Geographical Society*, January: Hobart.

McKenna, M., Roche, M., Le Heron, R. and Mansvelt, J. (1997b), 'Food Regimes Research in New Zealand's Apple Industry: Insights From the Orchard', *Proceedings of the International Geographers Union Meeting on Sustainable Rural Systems*, July 6-9: Armidale.

McKenna, M., Roche, M. and Le Heron, R. (1997c), 'Sustaining the Fruits of Labour: A Comparative Localities Analysis of the Integrated Fruit Production Programme in New Zealand's Apple Industry', *Journal of Rural Studies*, forthcoming.

McMichael, P. (ed.) (1994), *The Global Restructuring of Agro-Food Systems*, Cornell University Press: Ithaca.

McMichael, P. (1996), 'Globalisation: Myths and Realities', *Rural Sociology*, Vol. 61, No. 1, pp. 25-55.

Ministry of Agriculture (1993), *Sustainable Agriculture*, MAF Policy Position Paper 1, MAF Policy: Wellington.

Moran, W., Blunden, G. and Bradley, A. (1996a), 'Empowering Family Farms Through Cooperatives and Producer Marketing Boards', *Economic Geography*, Vol. 72, No. 2, pp. 161-177.

Moran, W., Blunden, G., Workman, M. and Bradley, A. (1996b), 'Family Farmers, Real Regulation, and the Experience of Food Regimes', *Journal of Rural Studies*, Vol. 12, No. 3, pp. 245-258.

New Zealand Apple and Pear Marketing Board (various years), *Annual Statistical Review*, 1970-1995, unpublished statistics.

Pipmark Grower News (1997a), New Zealand Apple and Pear Marketing Board, 20 June.

Pipmark Grower News (1997b), New Zealand Apple and Pear Marketing Board, September 9.

Whatmore, S. (1995), 'From Farming to Agribusiness: The Global Agro-food System', in Johnston, R., Taylor, R. and Watts, M. (eds), *Geographies of Global Change: Remapping the World in the Late Twentieth Century*, Blackwell: Oxford, pp. 36-49.

Wilson, R. (1996), *New Zealand Integrated Fruit Production*, consultancy report prepared by AgFirst Consultants: Hastings.

4 New Zealand's organic food exports: Current interpretations and new directions in research

HUGH CAMPBELL AND BRAD COOMBES

Introduction

One industry leader in New Zealand horticulture recently commented that there is a 'revolution' occurring in New Zealand's plant-based industries (Bull, 1997). This revolution could be identified by one significant moment in 1992 when the New Zealand Kiwifruit Marketing Board (NZKMB) admitted its conventionally-produced kiwifruit were compromised under food safety criteria in the marketplace, and that dramatic changes would be required to the way New Zealanders produce kiwifruit (Campbell *et al.*, 1997). The response of the NZKMB defined a path that many other exporting organisations are starting to follow. By 1997, the marketing board, now renamed as Zespri International Limited (ZIL), completed the transition to 100 percent production of kiwifruit under organic or integrated pest management (IPM) systems. Given that within four years the production for export of sweet corn, peas, squash, tomatoes, broccoli, wine and honey will approximate the Zespri pattern of two tiers - a peak organic tier and a middle level IPM tier - it seems that talk of a revolution was not premature.

Three questions that might be asked about this development are:

- why was organic food the first group of 'green' products to be developed for exporting?
- why are New Zealand-based companies engaging in this strategy?
- what kinds of theoretical explanation might be found within the context of global trends in food production and consumption?

This chapter will briefly sketch the way in which organic food exporting is leading a wider development in the 'greening' of food exports from New Zealand. It will then address the kinds of theoretical schema in which global trading in organic foods is being located by other researchers. An examination of organic food exporting suggests that none of the current theoretical explanations for the global trade in organics is entirely adequate, and that a new explanation must be sought within the ongoing crisis of the world trade in food and in the breakdown of Fordism. It is within this wider context of crisis that the strategy of 'greening' food exports in New Zealand can be understood, and the role of organic food within this wider situation can be clarified.

The New Zealand case

The first signs of a broad 'greening' of New Zealand food products emerged in the organic food sector between 1990 and 1995, when organic food was developed as a response to crises emerging in the export food sector in New Zealand. It is only since 1995 (and, in reality, only in one industry - kiwifruit) that a larger second tier of 'green' products has begun to emerge. Consequently, the historical development of 'green' food exports from New Zealand is rooted in the organic production sector. This section will therefore examine, first and foremost, the emergence of organic food exporting.

In 1990, New Zealand had a very low level of organic food production. A Ministry of Agriculture report (MAF, 1991) estimated that the total value of organic food traded in New Zealand in 1990 was US$750,000 (NZ$1.1 million), and this was almost entirely traded in the domestic market. Despite New Zealand having only a population of 3.5 million people, this was an extremely small segment of New Zealand's overall food production in that year and represented a low level of per capita consumption of organic food compared to European countries.

Since that time, however, organic food production had increased significantly. Initially, two large corporate entities - Wattie Frozen Foods (later to become Heinz-Wattie), and the then-Kiwifruit Marketing Board (now ZIL) - began to experiment with organic production systems in 1990. In the case of ZIL, this emerged in parallel to similar experimentation in IPM production systems. However, while Heinz-Wattie and ZIL were the most significant players in the early stages of this development and have continued to expand their output, recent membership of the newly formed Organic Products Exporting Group indicates that at least 24 other organisations are coming to join them. In 1995, organic food exports reached NZ$6 million. By June 1996 this had doubled to NZ$12.5 million and continuing growth thereafter meant that by June 1997, exports stood at NZ$23.5 million (Saunders *et al.*, 1997). Alongside this, the domestic market has also expanded, rising from NZ$1.1 million in 1990 to NZ$10.5 million in 1997, thereby alleviating the fear that an organic food exporting industry would destroy the local market for organic food (Saunders *et al.*, 1997).

There are several unusual features of this development. The first is the involvement of large corporate entities in establishing links to export markets, although it would be a mistake to suggest that this development was entirely driven by corporate players. In fact, the large corporates were late-comers to an industry that had existed for several decades and had its own characteristic institutional form. The organic agriculture movement in New Zealand arose from a wide coalition of interests among urban food consumers, lifestyle residents in peri-urban areas, European migrants to New Zealand in the 1950s and 1960s, and direct contact with the British Soil Association (Campbell, 1996a). By 1983 this loose coalition had become institutionalised under the New Zealand Biological Producers Council (later renamed Bio-Gro NZ), which now administers the Bio-Gro label. This label has become the more prominent form of organic certification, with the Biodynamic Steinerist group providing an alternative but less numerically significant label (Campbell, 1996a).

Another unusual feature of the developmenbt of the organic industry was that, unlike the situation in Europe, organic producers in New Zealand received little attention from the State, with no subsidies or incentive for growers to convert to organic production (and this still remains the case). Despite this, by 1990 the organic agriculture movement had transformed itself in three ways:

- by the institutionalisation of one main labeling system;
- by the formalisation of written standards for organic production, and;
- by the establishment of strong international links between New Zealand organic agriculture and international bodies like the International Federation of Organic Agricultural Movements (IFOAM).

This was the situation encountered by Wattie Frozen Foods and the NZKMB, whose organic programmes benefited from a number of existing features of the industry. First, organic techniques had already been successfully developed in the key regions and for the key commodities. Second, the labelling and certification systems were in place, albeit in an amateurish form. Third, New Zealand organic products and labels were already well recognised in overseas markets through the international links formed by Bio-Gro NZ.

While the situation that developed within the alternative agriculture movement provided one important context for the activities of large corporate firms, these entities were also responding to wider political and economic changes in New Zealand agriculture. The New Zealand economy, and agriculture in particular, were extensively deregulated after 1984. Both the macro- and micro-economic environments were altered, with the removal of subsidies and price supports, the restructuring of the Ministry of Agriculture and the removal of tariff barriers having particular bearing on the nature of food production and exportation. In such a highly deregulated policy climate, large corporate entities attempting to develop more genuinely 'green' products had a truncated range of alternatives. We contend that organics became a major vehicle for green product development principally because of the official neglect of any other 'low input' systems of agricultural production. Organics was the *fait accompli* of green development because existent practices had many advantages for capital over unproven developmental projects. The history of capitalist development suggests that business interests have far greater success in commodifying and appropriating partially developed production forms than in developing production systems and techniques independently.

The experience of the years from 1990 to 1996 was that large corporate entities provided the organic movement with a number of opportunities: first, immediate linkage to mature markets overseas - particularly in Japan; second, a vastly improved sense of legitimacy with both government and conventional farmers; and, third, an in-house research and development infrastructure for supporting organic development. In return, the large corporates capitalised on the labeling system, skills and 'green legitimacy' of the New Zealand organic movement

Given the trade-offs involved between the large corporates and the organic agriculture movement, this was obviously not an easy relationship. The arrival of the large corporate entities placed pressure on both the institutions behind the organic certification system and the ideological loyalty of many long term members of the organic movement (Campbell, 1996a). Their arrival did, however, increase the legitimacy of organic agriculture among many conventional growers, which was reflected in a relatively large number of conventional growers converting to organic production (Campbell, 1996a).

Perhaps the most interesting feature of this uneasy relationship was the establishment by the organic agriculture movement of a formal body of production standards. Prior to 1983, organic growers were declared to be legitimately 'organic' simply through their participation in the organic agriculture movement. In this context, large companies would not have been able to enter the organic scene and use this kind of legitimacy for their own products. By 1990, however, the development of a written set of standards for organic production and a formal inspection and certification service meant that any grower who met these criteria *had* to be certified, no matter what their ideological orientation towards sustainable agriculture. In this context, large companies could confidently contract with a number of recently-converted conventional producers, knowing that no matter how many misgivings the traditional members of the organic agriculture movement expressed, the recent converts could not be prevented from using the label if they met the technical requirements for organic production. While it is acknowledged that ideological adherence to the broader goals of organic agriculture remains part of the certification process, this has become harder to enforce. What had occurred was a transition from organic agriculture as a social movement committed to an overall goal of sustainability in New Zealand agriculture (a social movement which certainly still exists in residual form), to organic agriculture as a labelled quality able to be attached to certain types of food commodities. In the former situation, transnational corporations would have found no foothold; in the latter, organically produced food is able to be traded internationally like any other food product. The difference had been the formalisation between 1983 and 1990 of an 'objective' set of organic production standards.

This has had a mixed effect on the organic agriculture movement. There is no doubt that Bio-Gro NZ has developed a hegemonic position in the industry due to the success of the development of exporting and the revenues it generates. This success has, however, undermined some of the confidence of long-term adherents of the organic agriculture movement in the practices of Bio-Gro NZ. In 1994, a significant number of long term growers left the Bio-Gro NZ certification system due to increases in the inspection fees that Bio-Gro NZ had instituted in an attempt to professionalise inspection services (Saunders *et al.*, 1997). This signified a break with the past identity of Bio-Gro NZ as the representative of the organic agriculture movement in all its forms, and a move towards a new identity as the professional certifier of an internationally recognised set of organic standards.

In summary, the organic export industry is growing rapidly in New Zealand around a group of large corporate companies and a group of smaller exporters which is increasing in number. This development has revolved around a developing relationship between exporting companies and Bio-Gro

NZ to create a distinctive trade in certified organic products. The question remains, why has this happened? The exportation at the heart of this development suggests that the reasons must be found in either the target markets for these products or in the changing relationship between New Zealand and the global trade in food products. It comes as no surprise then, that the two significant attempts to understand global trading in organic foods seek answers in exactly these areas. The following section will briefly review these attempts to find a wider theoretical explanation for recent developments.

Theoretical interpretations of the global organic food trade

Having elaborated the evolution of organic exporting in New Zealand, it is now possible to examine the theoretical implications of this development. We might ask why organic agriculture has entered the theoretical narrative of the world food system at all when, in real terms it represents, at best, a marginal activity in the global trade in food?

To answer this question two theoretical positions must be addressed:

- organic food as propulsive in, or indicative of, an emergent third food regime;
- organic food as indicative of a new style of food network;

Organic food and a third food regime

Two prominent scholars of New Zealand's rural and food linkages to the world - Richard Le Heron and Michael Roche - have used our own research into organic agriculture (Campbell, 1996a; 1996b) as evidence to support their contention that a new style of production/consumption relationship is being forged in countries like New Zealand (Le Heron and Roche, 1996). Because of New Zealand's unfortunate national experiment in liberalised agriculture, transnational corporations (TNCs) are using it as a launching point to construct flexible supply systems of 'fresh' and 'green' products for lucrative markets in the first world. Their discussion to this point has strong parallels to Friedland (1994), and is certainly compatible with earlier renditions of our theoretical agenda (see Campbell, 1996b). They go further, however, by arguing that globalised TNCs are wedding flexible supply systems with discourses of sustainability (at a consumer and governmental level) to forge a new kind of global food linkage (Le Heron and Roche, 1995; 1996).

In the conclusion to their argument, Le Heron and Roche (1996) suggest that these new kinds of transnational delivery system may be indicative of an emergent third food regime. Three case studies are used to support this argument - a 'flexible' chilled meat supply company (Fortex), a tomato growing and processing company (Cedenco), and our own organic food exporting case study of Heinz-Wattie (referring, in particular, to Campbell, 1996b). Le Heron and Roche (1996) use these examples to suggest the emergence of a new global stability in 'new' food production sectors. However, in the two years since these cases were analysed, it has become clear that the conditions of these companies has been far from stable. Both Cedenco and Fortex have suffered major reversals in their New Zealand operations, and the fate of these two companies, as well as the ongoing uncertainty in the Heinz-Wattie case, highlights a lack of stability in 'new' food production sectors (Coombes, 1997), which may be characteristic of current food systems, but is somewhat incompatible with the notion of a stable food regime. A strict reading of Friedmann and McMichael (1989) would suggest that the notion of a historical global regime of food relations is, by its very nature, stable over a particular historical period.

The second main difficulty with the food regimes position is that it makes the transition from a few cases in New Zealand to the concept of a food regime rather rapidly. One should also point out that Harriet Friedmann uses very similar evidence of corporate involvement in the health food industry to argue exactly the opposite - that such moves are not indicative of the emergence of a third food regime, but of a desire by large food corporations to 'window-dress' their existing structures and retain the basic structures of the second food regime (Friedmann, 1993: 228; Campbell, 1996a).

Consequently, while it is possible that a third food regime may emerge in the future (see McMichael, forthcoming), the situation in New Zealand meets neither the spatial or temporal criteria of the accepted notion of a global food regime. In the case of organic food, a meso-level explanatory framework as exemplified by the food networks approach, might work just as well, if not better.

Organic food and food networks

The second use of organic food in recent theory is still only nascent, but occurs in an even higher profile forum of debate. Recent work by Terry Marsden and Alberto Arce (Arce and Marsden, 1993; Marsden and Arce, 1993; 1995) on the concept of the *food network* has had a significant impact on debates about the nature of global food trade. This meso-level concept brings to centre stage the subjective aspects of food systems, detailing the agency of a number of different parties in constructing the subjective 'quality' of food products. This excursion into a post-modernist vision of the world of food has certainly been provocative and suggests that a new phase of global food relations, in which the subjective constructions of food 'quality' will become increasingly prominent, is starting to develop. Yet again, organic food emerges as an exemplar of this new kind of food relationship by indicating the way that shared ideas and beliefs over healthy and 'ideologically sound' foods can form 'new consumption niches linking certain producers, retailers and consumers at a transnational scale' (Marsden and Arce, 1995: 1274).

This attempt to move towards subjectivity catches some of what makes the analysis of organic food so intriguing. Marsden and Arce are not alone in this. Two other contributors to the global food literature - Bill Friedland and Alison James - also link organic food consumption to 'subjective' factors in consumer purchasing by clearly situating the rise of organic food trading within consumer reactions to food scares and health concerns (Friedland, 1994; James, 1993).

As a meso-level explanatory concept, the idea of the food network introduces some interesting and important features to the analysis of organic food. Some prior contributors to the agricultural sustainability debate have tended to focus solely on an all-powerful vision of corporate actors who impose their will upon both farmer producers and end consumers (see Campbell, 1997). There is no doubt that the idea of the food network moves beyond such approaches by successfully identifying the multiple and negotiated relationships and the shared subjectivities involved in the production, distribution and consumption of organic food. In fact, it would be hard to improve on organic food trading as an example of a food network in operation (see Campbell, 1997).

The difficulty with such an excellent fit between this explanatory framework and organic food is that organic food is quite possibly very *atypical* of general global food commodities. There are no other groups of food products which have such a concretised set of shared discourses on a global scale. Once the focus moves away from organic food, the idea of the food network may end up not having such a high degree of theoretical utility.

A second point of concern is that, in principle, the idea of a food network implicitly gives equal explanatory weight to all participants in the network (even if, in practice, food network theorists have strongly favoured the consumer as the driving force behind networks). The New Zealand case demonstrates that such an approach might be deceptive. The food network concept is undoubtedly useful for understanding the process of alliance building (both structurally and subjectively) that took place between the organic agriculture movement in New Zealand, a number of conventional growers, government organisations, corporations, key distributors in markets like Japan and the US, and end consumers in these countries. However, the development of organic food exporting was most strongly influenced by the *corporate businesses* involved in the process. Prior to 1990, all the actors in the network (excepting the large corporates) were present and linked in some form, but the overall volume of organic food being produced and traded was trivial. It was the arrival of the large corporate players that expanded organic production, as noted earlier, from NZ$1.1 million in 1990 to NZ$23.5 million in 1997. Many commentators in New Zealand are concerned as to whether such an expansion was desirable or served the goal of creating sustainable agriculture. However, for the purposes of this discussion, this rapid expansion is *the* key feature which demands an explanation. Either the development of organic exporting should be explained as a network, which implicitly gives relatively even weighting to the contribution made by different parties to the network, or the development is seen as a predominantly corporate-driven phenomenon in which the arrival of corporate players transformed the organic industry from total insignificance to something demanding further analysis. The evidence from the New Zealand case suggests that we should adopt the latter view. Consequently, the motivations of the corporate players are very important and must be given more explanatory weight than the contributions of others participants in the organic food exporting network (without ignoring the other parties entirely).

This case for understanding the motivations of the corporate actors must be tempered with a crucial proviso. Campbell (1997) argued that a focus solely on corporate actors was ultimately deceptive and avoiding such reification was critical for understanding organic food exporting. Simply identifying a powerful corporate actor within a food network created a distorted view of global food trading. The corporate actors in the organic food network in New Zealand were engaged in organic food trade in direct synergy with their trade in conventional foods. Organic foods were linked as 'keyhole products' designed to assist the sale of conventional products in key markets. Furthermore, strategies to develop organic foods were not just determined by the existence of lucrative niche markets for organic foods, a view that would be totally compatible with food network theory; rather they were a response to the wider dynamics emerging in markets for conventional products. In particular, corporate motivations were influenced by the possibility of 'green protectionism' compromising the market prospects of conventional products.

By seeking to partition particular food networks from total food markets, the food network approach (and implicitly the wider commodity system approach) is open to the suggestion that such partitioning is a form of reification. The New Zealand case of organic food exporting clearly demonstrates this by showing both the advantages of a food network approach within limited parameters and the ultimate need to move beyond the network to understand why corporate actors are behaving in particular ways.

Green protectionism

While there is no room in this chapter to engage in a detailed discussion of the emerging trends in the political economy of global food markets which influenced organic food exporting, Campbell and Coombes (forthcoming) take this argument to its logical end-point. It is appropriate, therefore, to provide a brief indication of the kinds of theoretical explanation that are presented there.

The experience of large exporting entities in New Zealand, like Heinz-Wattie and the Kiwifruit Marketing Board, was that a new dynamic of 'green protectionism' was emerging in First World markets like Japan and Europe. This reflected the continued breakdown of the Fordist mode of growth, under which a system of border- and non-border tariffs emerged to protect national agricultural economies. During the prolonged failure of Fordism and with the rise of neo-liberalism in the present regulationist crisis (Lipietz, 1992), the legitimacy of this type of protection has been reduced. With no new mode of global regulation solidifying (Peck and Tickell, 1994b), but with farming lobbies in Europe, the US and Japan largely retaining their Fordist-style income supports (Grimwade, 1996; Paarlberg, 1997), governments in each region face serious fiscal crises over their agricultural policies. Green protectionism represents compensation for the reduction of tariffs in the post-Uruguay Round of the GATT, and even if this reduction has not proceeded with the speed or extent that was first expected (Paarlberg, 1997), environmental barriers are increasingly more salient to New Zealand's exporters.

There are three important trends in the development of green protectionism. First, there is a re-direction of price supports for core-nation producers towards 'environmental support' (Saunders *et al.*, 1997), which has given governments in New Zealand's destination markets a heightened sense of legitimacy in strengthening environmental non-tariff barriers. Second, there is the establishment of heightened sanitary and phytosanitary barriers for imported foodstuffs (Ingersent *et al.*, 1995). Third, there are much more stringent controls on pesticide residues. The results reported in Campbell (1996a) and Campbell *et al.* (1997) clearly demonstrate that concerns over green protectionism were a major factor in the decision making of corporate players contemplating organic production. What they acknowledge is that while organic food might be a useful niche market to cultivate, and has the potential to provide companies with lucrative returns in the short term, there is a far greater long-term benefit if organic production systems can be mainstreamed or successfully adapted to create low-residue products meeting 'food safety' requirements in First World markets. In other words, organic systems are either providing a product or creating systems that can be mainstreamed to products that will by-pass green protectionist barriers. Whatever the case, the growth in organic food exporting from New Zealand is synergistically linked to the environmental and health threats to conventional exporting, and these threats are in turn linked to the continuing contradictions of the last remnants of Fordism and its breakdown.

Conclusion

This chapter has sought to outline what is happening in organic agriculture in New Zealand and to briefly review some of the broader theoretical attempts to explain its development. In conclusion, our analysis has some sympathy with both the third food regime and food networks approaches. While our analysis shares a great deal with Le Heron and Roche's (1996) interpretation of events in New Zealand, we feel it is premature to conclude that a nascent global food regime is forming around such developments as 'fresh' and 'green' production by TNCs in a deregulated agriculture sector such as New Zealand's. It is more likely that if (rather than when) a globally operational food regime does emerge, its genesis will be in the protectionist politics of post-GATT first world markets rather than in the hopeful experiments in exporting countries like New Zealand, which are essentially a response to these trade developments. Similarly, the food network approach captures much of the essence of recent events in organic food exporting in New Zealand. It is likely, however, that organic food is *too* good an example of a food network in operation. For the idea of a food network to have greater weight as a meso-level theoretical framework, it needs to be applied to more mainstream aspects of the world food system.

By identifying the dynamic of 'green protectionism' as crucial to the actions of those large corporations that are driving the development of organic (and IPM product) exporting in New Zealand, a challenge to existing theory emerges. The apparent centrality of the shifting politics of global trade protection suggests that the food network approach is too reified, and the food regimes approaches appears unable to cope with a period characterised by instability rather than stability. Campbell and Coombes (forthcoming) contend that this crisis is part of the breakdown of Fordism, and that those theoreticians who coined the notion of Fordism - French Regulationists like Lipietz (1987; 1992) - have also provided the means necessary to understand the demise of Fordism and what that means for a food exporting nation like New Zealand. While it appears to be true that a 'revolution' is occurring in some key food exporting industries in New Zealand, the established canon of agri-food theory is having difficulty in accounting for it.

In conclusion, the emergence of organic food exporting in New Zealand since 1990 has been strongly influenced by political dynamics in trade protection. Organic food was the *fait accompli* of any wider 'greening' of New Zealand food products as it was the only green production system with an existing infrastructure for certification, established skills in production developed by pioneer growers in the organic agriculture movement, and a ready market in key niche markets. After five years in which organic agriculture has been predominant in the strategies of New Zealand exporters, this chapter suggests that a new phase is emerging in which organic forms only the top tier, with a second tier of IPM-style or low-input production providing the fastest growing group of 'green' foods being exported from New Zealand. While this chapter has concentrated mainly on organic foods, future research will have to embrace both organic and second tier 'green' foods (and the dynamics between them) if we are to understand the response of exporting organisations in New Zealand to the global politics of green protectionism.

References

Arce, A. and Marsden, T. (1993), 'The Social Construction of International Food: A New Research Agenda', *Economic Geography*, Vol. 69, pp. 293-311.

Boyer R. (1990), *The Regulation School: A Critical Introduction*, Columbia University Press: New York.

Bull, P. (1997), Zespri International Limited, personal communication.

Campbell, H. (1994), *Regulation and Crisis in New Zealand Agriculture: The Case of Ashburton County 1984-1992*, Unpublished PhD Thesis, Charles Sturt University: Wagga Wagga.

Campbell, H. (1996a), *Recent Developments in Organic Food Production in New Zealand: Part 1, Organic Food Exporting in Canterbury*, Studies in Rural Sustainability, Research Report No. 1, Department of Anthropology, University of Otago: Dunedin.

Campbell, H. (1996b), 'Organic Agriculture in New Zealand: Corporate Greening, Transnational Corporations and Sustainable Agriculture', in Burch, D., Rickson, R. and Lawrence, G. (eds), *Globalization and Agri-food Restructuring: Perspectives from the Australasia Region*, Avebury: Aldershot, pp. 153-169.

Campbell, H. (1997), 'Organic Food Exporting in New Zealand: the Emerging Relationship between Sustainable Agriculture, Corporate Agri-business and Globalising Food Networks', in De Haan H., Kasimis B., and Redclift M. (eds), *Sustainable Rural Development*, Ashgate: Aldershot.

Campbell, H., Fairweather, J. and Steven, D. (1997), *Recent Developments in Organic Food Production in New Zealand: Part 2, Kiwifruit in the Bay of Plenty*, Studies in Rural Sustainability, Research Report No. 2, Department of Anthropology, University of Otago: Dunedin.

Campbell, H. and Coombes, B. (forthcoming), 'Green Protectionism and Organic Food Exporting from New Zealand: Crisis Experiments in the Breakdown of Fordist Trade and Agricultural Policies', *Rural Sociology.*

Coombes, B. (1997), *Rurality, Culture and Local Economic Development*, unpublished PhD Thesis, University of Otago: Dunedin.

Friedland, W. (1994), 'The New Globalization: The Case of Fresh Produce', in Bonnano, A., Busch, L., Friedland, W., Gouveia, L. and Mingione, E. (eds), *From Columbus to ConAgra: The Globalisation of Agriculture and Food*, University Press of Kansas: Lawrence, pp. 210-231.

Friedmann, H. (1993), 'After Midas's Feast: Alternative Food Regimes for the Future', in Allen, P. (ed.), *Food for the Future: Conditions and Contradictions of Sustainability*, John Wiley and Sons: New York.

Goodman, D. and Watts, M. (1994), 'Reconfiguring the Rural or Fording the Divide? Capitalist Restructuring and the Global Agro-Food System', *Journal of Peasant Studies*, Vol. 22, No. 1, pp. 1-49.

Grimwade, N. (1996), 'Agricultural Protectionism', in *International Trade Policy - A Contemporary Analysis*, Routledge: London.

Ingersent, K., Rayner, A. and Hine, R. (1995), 'Ex-post Evaluation of the Uruguay Round Agriculture Agreement', *The World Economy*, Vol. 18, No. 6, pp. 707-728.

James, A. (1993), 'Eating Green(s): Discourses of Organic Food', in Milton, K. (ed.), *Environmentalism: The View from Anthropology*, Routledge: London, pp. 205-218.

Le Heron, R. (1989a), 'A Political Economy Perspective on the Expansion of New Zealand Livestock Farming, 1960 - 1984: Part One. Agricultural Policy', *Journal of Rural Studies*, Vol. 5, No. 1, pp. 17-32.

Le Heron, R. (1989b), 'A Political Economy Perspective on the Expansion of New Zealand Livestock Farming, 1960 - 1984: Part Two. Aggregate Farmer Responses - Evidence and Policy Implications', *Journal of Rural Studies*, Vol. 5, No. 1, pp. 33-43.

Le Heron, R. (1993), *Globalised Agriculture: Political Choice*, Pergamon Press: Oxford.

Le Heron, R. and Roche, M. (1995), 'A 'Fresh' Place in Food's Space', *Area*, Vol. 27, No. 1, pp. 23-33.

Le Heron, R. and Roche, M. (1996), 'Eco-Commodity Systems: Historical Geographies of Context, Articulation and Embeddedness Under Capitalism', in Burch, D., Rickson, R. and Lawrence, G. (eds), *Globalisation and Agri-food Restructuring: Perspectives from the Australasia Region*, Avebury: Aldershot, pp. 73-89.

Lipietz, A. (1987), *Mirages And Miracles: The Crises Of Global Fordism*, Verso: London.

Lipietz, A. (1992), *Towards a New Economic Order: Post-Fordism, Ecology and Democracy*, Oxford University Press: New York.

MAF (1991), 'A Proposed Policy on Organic Agriculture', *MAF Policy Position Paper No. 1*, Ministry of Agriculture and Fisheries: Wellington.

Marsden, T. and Arce, A. (1993), *Constructing Quality: Globalisation, the State and Food Circuits*, Globalization of Agriculture and Food Working Paper Series, University of Hull and the Agricultural University: Wageningen.

Marsden, T. and Arce, A. (1995), 'Constructing Quality: Emerging Food Networks in the Rural Transition', *Environment and Planning A*, Vol. 27, pp. 1261-1279.

McMichael, P. (forthcoming), 'Global/Regional Implications of the East Asian Food Import Complex', in *World Development*.

Paarlberg, R. (1997), 'Agricultural Policy Reform and the Uruguay Round: Synergistic Linkage in a Two-Level Game?', *International Organisation*, Vol. 51, No. 3, pp. 413-444.

Peck, J. and Tickell, A. (1994a), 'Searching for a New Institutional Fix: The After-Fordist Crisis and the Global-local Disorder', in Amin, A. (ed.), *Post-Fordism: A Reader*, Blackwell: Oxford, pp. 280-315.

Peck, J. and Tickell, A. (1994b), 'Jungle Law Breaks Out: Neoliberalism and Global-Local Disorder', *Area*, Vol. 26, No. 4, pp. 317-326.

Peck, J. and Tickell, A. (1995), 'The Social Regulation of Uneven Development: 'Regulatory Deficit', England's South-East and the Collapse of Thatcherism', *Environment and Planning A*, Vol. 27, pp. 15-40.

Saunders, C., Manhire, J., Campbell, H. and Fairweather, J. (1997), *Organic Farming in New Zealand: An Evaluation of the Current and Future Prospects of Organic Farming Including an Assessment of Research Needs and Capabilities*, Department of Economics and Marketing, Lincoln University: Canterbury.

5 The organic manifesto: Organic agriculture in the world food system

ANDY MONK

Introduction

In recent years the growth in the production of organic foodstuffs for both domestic and export markets has become increasingly significant in the agri-food systems of the developed economies (Hassall, 1990; 1995; Lampkin and Padel, 1994). This growth in volume and activity in the organic industry is in turn being mirrored in the activities of researchers and government bureaucrats. Many mainstream food industry players are also currently involved in monitoring such activities, and in some cases, larger corporations have entered organic markets for the purpose of establishing 'keyhole' market channels through which they can then move the bulk of their conventional produce (Campbell, 1996). Currently, a major supermarket chain in Australia is positioning itself to exploit this green image as a means of boosting its own conventional brand labels, through cross-promotional and associational advertising.

There is no doubt that attempts to capture a niche commodity are underway. But ironically this is bound to have two major effects. The first will be to catalyse further the expansion of production by the organic industry, both in terms of area and producer numbers. The second will be to expose more conventional food industry players to the world of organic food production and handling. Both these developments will catapult the industry to new levels of acceptance, while despatching the central tenets upon which the industry is based to all arenas of the food industry - including the hubs of innovation, the research and development cultures.

Significant players have entered into the field of organic primary production over the past few years. I will briefly outline two particular examples to illustrate this. The first involves developing-world producers who have sought organic certification by developed-country certification bodies as a means of establishing markets in more developed economies. The second example involves beef producers in semi-arid regions of Australia, who are linking up with an organic certification body, in order to demarcate their product in the market place as organic. Simultaneously, these producers are implementing a quality assurance (QA) scheme within their group, using the auditing and standards of the organic industry to effectively manage for quality. These developments can be seen as *core* effects that the organic industry is having on agricultural and food production practice.

These changes in agricultural practice have also resulted in chain reactions and downstream effects that are driving related changes in agriculture and food processing more generally. There are a number of areas where organic agriculture is moving out into the mainstream, and beyond the distinct boundary that is defined by the certification system for organic producers and those who deal directly in organic products. This is encouraging all those in agri-food production, from primary producers through to food factory managers, to look at their production practices in new ways. I will term these the *peripheral* effects of the organic industry, which are nonetheless very significant in terms of the changes they are imparting to mainstream agricultural and food industry practice. It is this mixture of core and peripheral effects which constitute the coming means by which the organic industry is to play a significant role in future moves toward more sustainable food production systems worldwide.

A quality assurance system with flesh

The Opal Beef Exporters (OBE) pastoral group range across an extensive area of Australia, from central to south central Queensland, to the border of the Simpson desert in south western Queensland, and down the Birdsville track into South Australia. This region is subsumed within the vast Lake Eyre catchment basin. There is a mixture of operations, involving cattle and sheep production, and cropping across highly diverse terrains, including seasonal channel country pastures, varied mulga country and sparse pebbly plains. This region is distinctive for its low incidence of cattle tick, buffalo fly, lice and internal parasites. As a consequence, there is less need for the extensive use of residual agrochemicals for control of these pests. Further, the lush natural seasonal pasture allows for extensive grazing for fattening on prime land. With few agricultural or other industries in this region, the OBE group enjoys the benefits of a relatively unpolluted, if historically overgrazed and ecologically modified, environment. While there are challenges in maintaining this type of production environment, the recognition by markets of the distinctiveness of the products from this region and the premium paid for its products, are viewed as strong incentives for growers to maintain the commitment to organic standards of production.

That such a group has aligned itself with the organic industry is in itself of interest since it represents a degree of legitimacy for the organic industry, which was absent only a few years ago. More importantly, it represents a recognition of the lengths to which conventional food producers are now willing to go to meet the requirements of a changing food market, which is demanding quality assured food products that are related to natural, if not clean, production environments. OBE perceives the organic certification system not just as a means of regulating its own members and establishing standards by which they allow new members into their marketing group, but also as a means of establishing a quality assured status for its products which incorporates minimal chemical residue control.

The issue of quality assurance (QA) is an interesting aspect of the ways in which organic practices and regulation have come to the fore in the food industry. QA status is now a significant arbiter in determining supplier relationships with major food establishments, while the export of food products is increasingly requiring QA certification as a matter of course. The organic industry has for some time now not only had an established form of QA in terms of the auditing of recording and management systems, but also has standards of management attached to these auditing requirements to which operators must conform. While QA and auditing systems have been around for some time, they have only very recently begun to be adopted in the Australian food production and market environment. Conventional QA is based upon auditing management practices and recording systems set by the individual requirements of clients and their operations. In this context there are no set standards for compliance beyond those established by the operator or group, and therefore there is no real sense in which farm management and land stewardship practice is necessarily regulated or modified for the better. Whether this addition of organic industry practices to the QA realm will be a feature of future developments in the agricultural and food industry sectors, remains to be seen. There appear to be some gradual moves in this direction with the establishment of such QA programs as Clipcare, Q-Care and Cattlecare, where operators are audited and assessed on a range of practices which lead to protection and possible labelling of the end product for sale. The establishment of a more comprehensive farm management program which is rigorously audited for compliance with a range of requirements, from environmental monitoring to pasture enhancement and stocking rate limits, is perhaps a long way off. The auditing and assessment which the organic industry exhibits stands as a model which future programmes may emulate. Such programmes focus on the production process itself and its impact upon the surrounding physical and technical production environment, and not just upon the end product.

The example of OBE, as with many large producers and processors looking into the expanding organic market, is an example of the core effect - that is, direct producer uptake of organic practices. It is having a direct impact on behaviour, as producers orient elements of their operations towards a changing market place. As this trend unfolds, these producers are being influenced by organic guidelines, and applying these practices to other areas of production. For instance, processors are realising the ease with which some organic practices can be superimposed upon conventional processing systems, in ways that eliminate once taken-for-granted chemical treatments for storage or preservation. Meanwhile, primary producers are asking questions regarding pest, disease and weed control, using modernised versions of earlier practices prior to residual agrochemical availability. These more peripheral effects do not rise only from an alignment with organic production, but there will increasingly be found reason and example within the organic set of practices for producers and processors to switch to what is termed 'cleaner and greener' production.

Organics in the Asian region

The involvement of the organic industry in developing countries is an interesting development, which is not so much concerned with introducing modifications to an agricultural region and its management practices, as it is with ensuring that traditional production systems only evolve along organic or sustainable lines into the future. During a period which is seeing major inroads into, and modification of, native forest systems and an influx of industrial agroforestry, this move is effectively helping to temper the influx of agrochemicals and sometimes inappropriate agro-industrial practices in these regions. These traditionally diverse production systems can often be highly productive whilst maintaining relatively high employment levels, where discerning 'ecological' consumers can be located for their products.

The National Association for Sustainable Agriculture, Australia, (NASAA) is one of a number of internationally accredited bodies which is inspecting and certifying producers in countries such as Sri Lanka, Indonesia, Papua New Guinea and Nepal, and in places such as Japan and Singapore, where indigenous certifying bodies do not yet exist, or are in their infancy. Many of these commercial moves have been part of the establishment of 'fair-trade', whereby developing country products are sourced and sold into developed country markets, where there exists a niche demand to support products embodying organic production practices. The move by European-type organic agriculture certifying bodies into these distinctly different production environments is interesting in that there has been a cultural exchange of views on the way to establish and assess the sustainability of these types of production systems. Nonetheless, standards have been established which come under the rubric of 'organic' and producers are being encouraged, as are certified Australian producers, to adhere to these standards to gain a market advantage. Primary producers often supply one of a number of major companies involved in this trade. These companies - sometimes indigenous, other times foreign - act as the banner under which producers are encouraged to congregate. This allows many small holders, sometimes engaged in projects which involve whole villages, to gain an advantage through collective marketing.

Certified organic products from these regions include traditionally produced tea, spices, tropical fruits, nuts and vegetables and even certified timber products. The production environments range from highland tea plantations, through biologically diverse forestry systems, down to lowland intercropping of mixed vegetable and fruit production. There are requirements for producers to comply with standards for composting and the use of animal manures and other natural fertilising inputs, through to establishing or maintaining biologically diverse production systems, logging on a non-clearfell basis and re-establishing longer term timber and tree crop varieties. There is also a requirement that operators contribute to the region's socio-economic development based upon the premium that is enjoyed from niche export markets. Products are aligned with marketing strategies that position them as environmentally sound and socially equitable.

What has been noted by producers, as well as the organisations marketing these products, is the impact such practices are having on the region in which they are based. There are a number of neighbouring producers who are emulating the practices of these certified operators. Where whole villages have been certified, those in neighbouring areas have copied some of the practices of the certified operators. This has included composting and mulching techniques; establishment of trenches for entrapment of water (as well as other practices associated with sloping agricultural lands technology - SALT); intercropping with a variety of species; planting in ways which encourage beneficial relationships between species; introducing exotic varieties which have markets both domestically and for export, and so on. Again, while there is a direct (core) impact on those operators who come under the certification banner, with sometimes considerable pressure placed upon them to meet those standards, there is also the downstream (peripheral) effect where these practices seep out into the neighbouring environment.

Agri-cultural capital

These changes need to be seen as more than merely mediated and controlled by direct market forces. While there is no doubt that organic production is strongly catalysed by growth in demand for organic and clean and green products, there are cultural shifts taking place which surround and add complexity to these markets. Countries of the European Union (EU), where there have been significant subsidies and benefits made available for both conversion to, and production within, organic production regimes, show key aspects of this change. Support has also been made available in some cases for developing-country agricultural projects that have entered the fair-trade and organic markets in the developed world. The 'semiperipheral' nations such as Australia and New Zealand have not exhibited this degree of change, but rather have moved in the other direction, i.e. the removal of subsidies and the adoption of policies of 'economic liberalism' (Lawrence and Vanclay, 1994). Nonetheless, while expanding organic markets are focussing the attention of once-conventional marketers, it is becoming at the same time culturally more acceptable to acknowledge organic ideas and practices. While it is still acceptable to castigate the cantankerous and impossible demands of the organic standards, it is more difficult to disparage its potential for growth, or to question its efficacy at a production level.

The notion of cultural capital as outlined by Pierre Bourdieu helps highlight some of the ways in which the organic markets have been mediated by a cultural change (Bourdieu, 1990). In Australia in the early 1990s, it was socially acceptable (and one gained more cultural capital) by either denigrating organic agricultural ideas or promulgating the notion that it was impossible that organic production would ever be anything other than a niche practice. Now, however, there is a growing realisation among a range of conventional primary producers, and more importantly major food processors and retailers, that this market is developing ahead of their ability to capture it in advance of their competitors on either the domestic or international scene. While the market may remain a niche, it is becoming a more attractive market for consumers by being positioned to represent 'naturalness' and 'clean' and 'green products'. Modern food markets are increasingly being constituted by a splintering and scattering of such niche market products (Senauer *et al.*, 1991). What defines the organic industry is its adherence to stand-alone production guidelines and regulated practices, which makes this particular niche market more than merely a product market. It is also a process that represents a clear divergence from mainstream food practices.

The cultural aspects of this change in practice are complex and somewhat contradictory, particularly given the contrary and fickle nature of consumer markets and their consequent impact on producer behaviour. For instance, there is often a great disparity between measured attitudes and ultimate buying behaviour of consumers for clean and green products. In a study conducted by Parigi and Clarke (1994), recent consumer attitudes were compared to those in the late 1980s, which revealed a significant decline in the number of consumers who believed that chemical residues in foods were a major concern. However, this has also come at a time when consumer buying behaviour in Australia has pushed both sales and production of organic product up. While exports are the main impetus behind the organic industry to date, most developed countries are also experiencing a rise in domestic demand and consumption.

This may well be a market fad that will pass with time. What is less likely to fade so quickly is the legitimation crisis of conventional agriculture, resulting from land degradation, chemical residues in end products, over-utilised water resources, the impact of drought, and a market setting which is squeezing producers to diversify and/or to raise their level of production ever higher. The consequence of this is that there is considerable pressure to find solutions to these seemingly intractable problems, of which organic agriculture offers more than one avenue of relief. The 'hype' on organic agriculture may well be just that, but the real issue is whether this enthusiasm will lead to overshooting the mark on what is economically and technically practical within the agricultural sector in the longer term.

What role will organic agriculture play in sustainable agriculture?

There is a range of benefits which organic agriculture can bestow upon conventional culture, at a production, regulation, research and market level. It has been argued elsewhere that organic agriculture is a paradigm distinct from conventional agriculture (Wynen, 1996). This view however, is made complex by significant elements of convergence between organic and conventional interests (Dumaresq *et al.*, 1996; Monk, 1998). This cultural and technical spill over is the most crucial of these, but paradoxically, this will only continue if organic and related research continues to push into areas which are paradigmatically different, and if it continues to protect its core. This will have a number of effects on the broader uptake of an essential core of principles and practices of organic agriculture, rather than some diluted versions. There is a need to see that organic agricultural practice will continue and develop on a number of planes - one at a core level which adheres directly to nationally or internationally regulated standards of production, and another at the peripheral level where there is partial emulation of these core practices by outsiders. Similarly, organic agricultural research will function in two ways - core or pure organic related research, and the effects of that research which may be peripheral to conventional interests but has significant potential for uptake by a wide range of producers in piecemeal fashion. The impacts of a future organic movement are therefore varied.

Production and regulation

At a direct level, it is clear that primary producer adherence to organic agricultural standards opens an avenue for regulated protection of land and water resources. Producers remain certified within the organic agricultural system as long as they adhere to the often rigorous demands of organic production. These demands may include requirements for the building up of soil fertility and humus, managing stock to prevent disease or pest infestation, and preventative environmental strategies aimed at capping water use requirements, and minimising land degradation and soil erosion. It should be noted that generally, while Landcare and similar projects are aimed at establishing land protection and best practice for land-stewardship, there is no pressure quite like the organic certification system, in its mixture of market and regulatory forces, to maintain particular production practices.

For organic systems therefore, there is scope for regulation of wildlife protection, enhancement of native scrub and pastures, capping water usage rates and practices and stocking rates, as well as the encouragement of measures aimed at preventing or reducing salinity, off-farm pollution, and a range of other preventative and palliative measures. As is occurring in Sri Lanka,

Indonesia and other developing countries, there is also growing scope to help protect biologically- and culturally-diverse production systems, through direct market incentives to remain in the certification system. The production of chemical-free beef in the channel country of Queensland in Australia has also catalysed support for the prevention of the influx of cotton production into regions that ultimately flow into the Lake Eyre basin. It has been argued convincingly by scientists that cotton production and its associated practices would pose salinity problems in this region as well as introducing residual agrochemicals into an otherwise 'clean' environment. There is no doubt, however, that the presence of these now-certified beef producers, who have an interest in maintaining an agriculturally clean environment for production and export purposes, has worked to consolidate a networked alliance to safeguard their position. In this way, those certified under organic standards come to change their practices and/or have an interest in adopting new land protection and enhancement measures. They also have a direct political and technical impact on their surroundings. Those who do not meet the standards are refused certification, while others who see a benefit to their enterprises invest the time and energy required to bring their activites in line with standards. There is an incentive for them to maintain these standards irrespective of other market pressures which might encourage them to overstock or crop at levels which could be deemed unsustainable in the longer term.

However, while conventional producers or those not operating within the organic regulatory framework have no direct dividends accruing to them through adherence to organic standards, there is a degree to which organic practices are seeping into the conventional farming world. This can be seen in the case of Sri Lankan producers located adjacent to or near organically-certified producers, who observe that organic practices are proving to give better returns to the farm, and can stabilise farms in times of drought while circumventing otherwise intractable problems with pests, diseases or weeds. For example, general conditioning and balancing of nutrients in soils helps to maintain health and disease prevention in crops and stock. Many conventional neighbours of organic producers in Sri Lanka are beginning to ask questions regarding such outcomes. Some, in turn, are trialing variations of these practices on their own properties. This 'demonstration effect' could be seen as marginal, but nonetheless it is a critical addition to the impacts which organic production will continue to have on conventional producers. It is also gradually opening the eyes of researchers to the potential benefits of certain organic management practices.

Most importantly, the establishment of regulatory schemes which are linked to market incentives are proving successful as a way of changing producer behaviour. The organic industry represents a practiced, if extreme, model of quasi-market regulation of land stewardship. Other programs within the conventional industry, such as the Kiwi Green programme in New Zealand (Campbell *et al.*, 1997), or the quality programmes of the wool and beef industries in Australia, are examples of these market-linked regulation programmes. Regulation of such practices as wildlife protection, land management and water conservation is yet to be imposed within such schemes. The existing organic model offers opportunities here to achieve real protection and best practice in land stewardship. It also allows for the existence and consolidation of subject matter for researchers looking into biological and more integrated on-farm 'ecological' research.

Research

The realisation of expanding export markets for organic products has generated a degree of government interest, and has resulted in the direction of resources into organic industry research and development. While these resources have been limited, the fact that they exist at all is of interest. This again is creating both a core effect as well as a peripheral one. Core effects come from larger operators or government bodies investing in research to directly benefit their (new) interests in organic production. The core effect is connected to the fact that organic production systems often push to the extreme the requirement for monitoring and biological control, while requiring a far more integrated, whole farm, approach to research. An example of this is integrated pest management (IPM) which, in the conventional field, still relies upon chemical intervention where this is deemed economically necessary. With a prohibition on agrochemical use, organic operators are forced to search for other solutions to their problems. In an organic orchard setting, the IPM monitoring and prevention of widespread pest infestation relies upon extensive pest and predator population knowledge and pushes the bounds of understanding and skill in the areas of biological and mechanical control.

This search, in turn, provides practical knowledge and skill in areas of IPM through catalysing the rate of research into areas that may not otherwise have been broached. This knowledge is then available for conventional use. Similarly, the requirement for minimal-to-zero residues in meat and other products has led to a cross-fertilisation in the areas of processing as well as primary production. Carbon dioxide gassing, steam, or chilling treatments are all now regularly practised in the wider food industry, while organic production methods at the farm level are being investigated for their efficacy in supplying solutions to the requirements of nil-or-low residue foods. While these changes have not always been inspired by the organics movement, there is a clear connection between workable, feasible models that can be drawn upon, and future technology uptake across the food industry; this is the fastest form of technology diffusion. The organic industry provides a number of arenas for developing and assessing these practices and technologies.

It has been noted that there is a paradigmatic difference between organic and conventional research requirements. With the confluence of interests involved in establishing cleaner and greener production and marketing systems, this paradigm separation is narrowing and, arguably in some cases, the two are indistinguishable. However, the organic industry's distinctive contribution is that it will continue to play an effective role in the wider food industry, by encouraging an ongoing search for both cleaner and greener production, both in terms of product and process. Its culturally confounding distinction with mainstream practice needs to be seen in this light - as a radical and unsettling influence that extends research boundaries and changes on-farm practices - a crucial element of any healthy research agenda.

Conclusion

One could argue that without specially-directed subsidies, the organic industry will be limited in its ability to meet the potential market demands for exports from Australia. No doubt, the industry will grow irrespective of the presence of subsidies and support schemes. But the industry will be both limited in its direct production ability and, more importantly, in its potential impact on the majority of producers who are peripheral to the organic industry. This would be a loss both from the point of view of the potential for world's best practice in agriculture, as well as leading edge research, which may lay claim to being on the path towards environmentally and economically sustainable agriculture.

References

Bourdieu, P (1990), *The Logic of Practice*, Stanford University Press: Stanford.
Campbell, H (1996), 'Organic Agriculture in New Zealand: Corporate Greening, Transnational Corporations and Sustainable Agriculture', in Burch, D., Rickson, R. and Lawrence, G. (eds), *Globalization and Agri-Food Restructuring: Perspectives from the Australasia Region*, Avebury: Aldershot, pp. 153-172.

Campbell, H., Fairweather, J. and Steven, D. (1997), *Recent Developments in Organic Food Production in New Zealand, Part 2, Kiwifruit in the Bay of Plenty*, Department of Anthropology, University of Otago: Dunedin.

Dumaresq, D., Greene, R. and Derrick, J. (eds) (1996), *Discussion Paper on Organic Agriculture in Australia*, Rural Industries Research and Development Corporation: Canberra.

Hassall and Associates (1990), *The Market for Australian Produced Organic Food*, Rural Industries Research and Development Corporation: Canberra.

Hassall and Associates (1995), *The Domestic Market for Australian Organic Produce - An Update*, Rural Industries Research and Development Corporation: Melbourne.

Lampkin, N. and Padel, S. (eds) (1994), *The Economics of Organic Farming: An International Perspective*, CAB International: Wallingford, UK.

Lawrence, G. and Vanclay, F. (1994), 'Agricultural Change in the Semi-Periphery: The Murray-Darling Basin, Australia', in McMichael, P. (ed.), *The Global Restructuring of Agro-Food Systems*, Cornell University Press: Ithaca, pp 76-103.

Monk, A. (1998), 'The Australian Organic Basket and the Global Supermarket', in Burch, D., Lawrence, G., Rickson, R. and Goss, J. (eds), *Australasian Food and Farming in a Globalised Economy: Recent Developments and Future Prospects*, Monash Publications in Geography No. 50, Monash University: Melbourne.

Parigi, P. and Clarke, R. (1994), *Consumer Attitudes, Perceptions and Behaviour with Respect to Chemicals in Fresh Food Production*, Agriculture Victoria: Melbourne.

Senauer, B., Asp, E. and Kinsey, J. (eds) (1991), *Food Trends and the Changing Consumer*, Eagan Press: St. Paul, Minnesota.

Wynen, E. (1996), *Research Implications of a Paradigm Shift in Agriculture: The Case of Organic Farming*, Centre for Resource and Environmental Studies: Canberra.

6 An end to Fordist food? Economic crisis and the fast food sector in Southeast Asia

DAVID BURCH AND JASPER GOSS

> At first we thought that the crisis would be mild and prices would
> only have to go up by 20 percent or so, but when the rupiah hit
> 11,000 to the US dollar, a loss of about 75 percent in six months, we
> could no longer even calculate price[s].
> (Steve Sondakh, Director of Hero Group Supermarkets, Indonesia,
> February 1998)

Introduction

From the late 1980s through to the late 1990s, many countries within Southeast Asia experienced high rates of economic growth.[1] Concomitant with this process was the emergence of a rapidly-expanding class of urban dwellers whose tastes and increasing purchasing power created markets for the products of international and local fast food chains. During this period, fast food companies such as McDonald's, Pizza Hut and Kentucky Fried Chicken (KFC) opened hundreds of outlets through East and Southeast Asia, responding to growth in the region and market saturation in the United States and other 'advanced' countries. By 1997, McDonald's had over 400 outlets in the ASEAN region (*Bangkok Post*, 11 March 1997). At the same time a number of local companies, such as the Jollibee Corporation of the Philippines, provided competition to the major North American chains, usually by offering variations to standard products in order to cater more effectively to local tastes.

Until very recently, most of the analysis of East Asian development tended to assume a continuing process of economic growth. In a similar vein, the market for fast food products was expected to grow. McDonald's, for example, planned to open 2,400 new outlets throughout the world in 1997, with 80 percent of these located outside the United States (*Courier-Mail*, 19 July 1997).

Clearly this is a significant development, but little attention has been paid to the local impacts of this spread of western-style fast food to Southeast Asia and the less developed countries, more generally. What work has been done has tended to focus either on the implied decline in nutritional standards as western fast food came to supplant local alternatives (e.g. food hawkers in Southeast Asia), or on the extent of cultural transformation associated with the spread of western consumption patterns, of which fast food is a symbol (Barnett and Cavanagh, 1994; see also Watson, 1997, for a recent discussion of some of these issues). Equally, the viability of the wider systems of production which emerged to support the expansion of fast food chains have largely been seen to be unproblematical in politico-economic terms. The collapse in 1997 of currencies in East and Southeast Asia and the subsequent downturn of a number of national economies (in particular Indonesia, Thailand, Malaysia and the Philippines), has raised questions about the continued viability of the models of development which these countries embraced.

This chapter examines the implications of the 1997-8 currency crisis from the perspective of the fast food industry in Southeast Asia, and attempts to evaluate some of the impacts this has had, or is likely to have, on local production and processing sectors. We believe that the analysis of the fast food sector in a period of crisis is also important in terms of our understanding of the major theoretical models which purport to explain changes in the agri-food system, and provides insight into the implications of the current crisis for the dominant model of development.

Of particular interest here is what the current crisis tells us about the notion of Fordism, which has in recent years been used to conceptualise the organisational strategies associated with the transformation of social orders, production systems and consumption throughout industrialised countries. Following from this, and by use of a number of case studies, we consider the argument that Fordist fast food in Southeast Asia coincided with a specific historical moment and that the crisis, currently occurring, problematises the continuity and stability of the fast food orders engendered by a previous era of capitalist development. Finally, the chapter comments on the ramifications of such processes for food security in countries which experience problems in supplying sufficient staple foodstuffs to the majority of their populations.

Fordism, fast food and the East Asian 'miracle'

Fordism, as an organising principle for production, has long been recognised as a revolutionary concept which separated twentieth-century manufacturing from previous eras of capitalism. The expansion of Fordist systems of production was promoted and encouraged through international business and multilateral institutions such as the IMF and the World Bank, during the decades after the Second World War. While the use of Fordism as an organising tool of social analysis is a relatively new concept, it has played an important role in social theory since the late 1970s, following the development of regulation theory. Those associated with the regulation school drew their theoretical roots, in part, from the Italian Marxist writer Antonio Gramsci (see Gramsci, 1971), and sought to explain the longevity of capitalist societies which were able to persist in spite of the consequences associated with inherent economic crises (Aglietta, 1979; Lipietz, 1982). For regulation theorists, Fordism as a theoretical concept was more than simply a system of production; it included whole networks of social relations, institutions and ideologies that formed in relation to production systems in industrialised capitalist societies (Hoogvelt, 1997). Fordism became an integral concept linked to notions of corporatism, Keynesian deficit financing, state intervention, import substitution and a mass consumer society where, to put it crudely, workers traded the 'militant' goal of collective ownership for wage increases linked to productivity and, in certain countries (e.g. the then-West Germany), limited participation in decision-making within industrial structures (Harvey, 1990; Jessop, 1990). Thus, what characterises a Fordist social order is, in fact, an historical moment when a *rapprochement* is achieved amongst powerful social forces which might otherwise be in conflict. Whether that moment has finished, with the industrialised countries having moved into a 'post-Fordist' era, is a matter of intense debate (for a detailed survey of this discussion see Amin, 1994).

This debate revolves around a number of themes related to changes in world orders over the last two decades. Firstly, during the period between 1945 and the early 1970s, there existed strong internally-led growth throughout most countries in the world, a situation consistent with a Fordist social order. By the late 1970s and throughout much of the 1980s, most countries in the world (with the exception of those in the East Asia region) experienced a decline in domestic rates of growth and an increase in levels of unemployment (Harvey, 1990). Second, Keynesianism as a means of economic management ceased to be a central plank of public policy, as it came to be increasingly challenged by monetarist policies which, among other things, led to reductions in the levels of public ownership, decreases in welfare spending and the removal of protective barriers for domestic industries (Teeple, 1995). Third, the use of increasingly sophisticated technological inputs, specifically computers, led to greater flexibility and reduced labour inputs (Lawrence and Vanclay, 1994). This, in turn, saw the rise of so-called systems of flexible accumulation, which were characterised by niche (as opposed to mass) markets, the growth of service industries (and an associated decline in manufacturing), specialised production with an increasing reliance on out-sourcing and sub-contracting (as against a dependence on economies of scale), and the rise of non-specific managerialism (in contrast to industry and firm specific management) (Lash and Urry, 1987; Harvey, 1990). As Arrighi (1994: 2-3) notes,

> organised capitalism [is]…jeopardised by an increasing spatial and functional deconcentration and decentralisation of corporate powers, which leaves processes of capital accumulation in a state of seemingly irremediable 'disorganisation'.

These factors constitute what has become known as a post-Fordist regime of accumulation. However, it has to be noted that this conceptualisation has not been universally accepted, and numerous problems remain, especially in the context of agriculture and food, and with reference to developing countries. In terms of the latter, Lipietz (1982) has argued that the extension of Fordism to the 'periphery' was a qualitatively different experience from that undergone in the industrialised countries, in part because wage-labour systems often had to be created, as opposed to being adjusted, as was the case in the First World. Further, the idea that there is a simple analytical and binary divide between Fordism and post-Fordism has been challenged on a number of fronts. Fine *et al.* (1996: 51) note this conceptual apparatus rests upon, 'ideal types of industrial organisation into which a much more complex and varied range of empirical possibilities do not adequately fit.' Additionally, the notion that agriculture worldwide should replicate industrial developments in both the First and Third Worlds has been strongly criticised (Goodman *et al.*, 1987; Goodman and Watts, 1994).

Fast food as Fordist food

The interesting point concerning fast food is that it *does* bear a remarkable resemblance to the ideal type industry of regulation theory (e.g. automobile manufacturing) in that the end-product is a mass-produced standardised commodity which allows few, if any, customer-induced variations, which is produced from inputs which are highly specified and are the same throughout the world, and which is able to exploit economies of scale in the production of a cheap and widely accessible product.[2] Equally, it seems clear that the 'older' segments of the fast food industry in North America (e.g. KFC, McDonald's and Pizza Hut) are approaching the point at which a transition to the more flexible modes of accumulation associated with post-Fordism, might be deemed necessary. Mathews (1989), for example, suggested that the crisis of Fordism in the mid-1970s was, in part, a consequence of western consumers being unable to continue to absorb the outputs of Fordist production systems. Consumers in the First World, could only utilise so many cars, washing machines, refrigerators and so on. For the system to continue to grow, it became necessary (among other things) to produce other, differentiated goods for niche markets and/or to penetrate markets in the less-developed countries for the mass-produced goods made available by advanced production systems.

In terms of the fast food sector, companies such as McDonald's chose the second option. Instead of a transition to a regime of accumulation characterised by flexible production of specialised fast foods for niche markets, most existing fast food companies responded to stagnating home markets by expanding rapidly into new international sites. Moreover, as this decline occurred, the countries of the East Asia region experienced significant growth in sectors traditionally associated with Fordist regimes, viz. manufacturing for mass markets (see Table 6.1), which gave rise to the conditions under which the demand for Fordist food within the region could grow.

This raises the possibility that what Lipietz (1987: 78) calls 'peripheral Fordism', might be a useful concept in situating the emergence of the fast food industries in Southeast Asia over the last two decades. Like Fordism in core countries, peripheral Fordism,

> involves both mechanisation and a combination of intensive accumulation and a growing market for consumer durables...[However] its markets represent a specific...[integration] of consumption by the local middle classes, with workers in Fordist sectors having limited access to consumer durables, and exports of cheap manufactures to the [core countries] (Lipietz, 1987: 79).

It is important to note an important qualification which is presented by the case of fast food. Unlike the manufacturing model, the emergence of peripheral Fordism in fast food does not necessarily imply the decline of mass markets for fast food in the industrialised centres. Indeed, the growth of mass consumption fast food in new markets may have underwritten the continuation of production systems dependent on economies of scale in the industrialised world. For example, greatly increased exports of frozen french fries from the United States to East Asia may serve to ensure the continuation of cheap supplies of this commodity to the domestic fast food sector in the United States.

At this stage, such issues remain unresolved, and in order to understand current changes it is necessary to examine the recent history and circumstances of the countries in Southeast Asia.

Transformations in Southeast Asia 1980-1997

According to the World Bank (1997: 234-5), in the period from 1980 through to 1995 the newly industrialising countries (NICs) of East Asia achieved average annual growth in gross domestic product (GDP) of 7.3 percent.[3] In contrast, the high income economies, encompassing most of the OECD, saw average annual GDP growth of 2.6 percent (World Bank, 1997: 235). For the period from 1965 to 1990, the East Asian NICs share of world exports rose from 3 percent to 9.1 percent, in marked contrast to developing countries as a whole, whose share in world exports fell by a fifth - from 24.2 percent in 1965 to 19.8 percent in 1990 (World Bank, 1993: 38). Yet, despite their apparent similarities, the NICs do not necessarily form a unitary or homogeneous category. There is a degree of divergence between the Northeast and Southeast Asian NICs, with the city states of Hong Kong and Singapore standing in marked contrast to others, in terms of sectoral contributions to GDP (see Table 6.1).

These differences find clearer expression when the percentage of the labour within given sectors is considered (see Table 6.2). Indonesia and Thailand still retain very large agricultural populations, while South Korea and Taiwan are more urbanised. These differences in levels of urbanisation, combined with the spending powers of different elements of the population, play an important role in the emergence of the fast food sector in Southeast Asia. In an industry dependent upon an expanding base of urban consumers with increasing incomes, sectoral labour shifts of one or two percent can amount to one or two million new urban dwellers in the case of countries such as Thailand and Indonesia. Thus, the growth rates of the 1980s and 1990s provided ideal conditions for the expansion of the fast food industry in Southeast Asia. (As will be seen later, the converse has occurred more recently, as the 1997 currency crisis resulted in falling incomes, growing unemployment and a return to rural areas by former urban workers).

Table 6.1   Sectoral contribution to GDP, various countries 1980-1995

	Agriculture		Industry		Manufacturing		Services	
	1980	1995	1980	1995	1980	1995	1980	1995
Singapore	1	-	7	9	29	27	62	64
Hong Kong	1	-	8	*8*	24	*9*	67	*83*
Indonesia	24	17	29	18	13	24	34	41
Thailand	23	11	7	11	22	29	48	49
Malaysia	22	13	17	10	21	33	40	44
Philippines	25	22	13	9	26	23	36	46
South Korea	15	7	11	*16*	29	*27*	45	*50*
Australia	5	*3*	17	*13*	19	*15*	57	*70*
Japan	4	*2*	13	*14*	29	*24*	54	*60*
USA	3	*2*	12	*8*	22	*18*	64	*72*

Source: World Bank (1997).
Figures in italics are for years other than those specified.

Table 6.2   Percentage of labour force by sector, various countries 1980-1990

	Agriculture		Industry		Services	
	1980	1990	1980	1990	1980	1990
Singapore	2	-	44	36	52	64
Hong Kong	1	1	50	37	49	52
Indonesia	59	57	12	14	29	29
Thailand	71	64	10	14	19	22
Malaysia	41	27	19	23	40	50
Philippines	52	45	15	15	33	40
South Korea	37	18	27	35	36	47
Australia	6	5	32	26	62	69
Japan	11	7	35	34	55	59
USA	3	3	31	28	66	69

Source: World Bank (1997).

The emergence of fast food in Southeast Asia

The major North American fast food chains started to expand out of their home base in the late 1960s and the 1970s, moving first into Europe, Australia and New Zealand. The industry in Southeast Asia showed some initial activity at this time, which was closely related to the presence of US troops in a number of countries of the region. McDonald's opened its first Asian outlets in 1971 in Japan and Guam, whilst South Korea has long had numerous well-established Pizza Hut outlets (*Bangkok Post*, 15 May 1995; *The Australian*, 14 April 1998). One of the earliest of the less-developed countries in the region to be targeted was the Philippines, where the KFC group opened its first outlet in Manila in the late 1960s. This initial venture failed and the company closed its operations in the Philippines and did not return until 1977. The KFC group was also the first company to enter the Malaysian market in 1975, preceding McDonald's by eight years (*Malaysian Business*, 1-15 January 1992: 14; 16 June 1994: 15). Subsequently KFC, McDonald's and Pizza Hut expanded rapidly throughout Asia, moving into Taiwan, the Philippines, Thailand, Indonesia and more latterly expanding into China and India. These major companies were followed by a number of smaller chains, such as Shakey's Pizza, and by 'home-grown' variants such as the Jollibee Corporation of the Philippines.

The spread of the major fast food chains can be explained by reference to what might be termed 'push' and 'pull' factors. 'Push' factors include the saturation of markets in developed counties, where despite attempts to increase sales and capacity utilisation (e.g. by marketing breakfast meals outside 'traditional' business hours), difficulties continue to be encountered. In October 1996, for example, McDonald's reported a decline in US domestic sales (excluding new restaurants) for the fifth successive quarter (*Asia Franchise and Business Opportunities,* April-June 1997). With over 12,000 McDonald's restaurants in operation in 1998 continued expansion is reducing the viability of existing operations, and 'franchisees have complained that the market is so saturated that new stores eat into [other stores'] profits' (*The Australian*, 14 April 1998).

As a consequence, McDonald's has come to focus on overseas markets for growth. By 1997, McDonald's operated more than 21,000 restaurants in 104 countries, and 4,000 of these were to be found in 17 Asian countries. It was in 1994 that the company came to rely on overseas markets for the largest share of its operating income, and by 1996, the ratio of overseas to domestic contribution to the company's operating income was in the order of 3:2 (*The Australian*, 15 October 1997; 14 April 1998).

In terms of the 'pull' factors, the expansion of the fast food industry in East and Southeast Asia was only possible where there emerged an effective level of demand for fast food products, which in turn assumed the emergence of a class of urban-based consumers who had the resources and the disposable incomes necessary to purchase such products. Certainly, such a class emerged in the newly-industrialising countries of the region, which were the focus of much of the foreign investment and technology transfer which accompanied the transition to more intensive modes of production.

Case studies of fast food in Southeast Asia

The case studies below examine the growth of fast food sectors within a number of countries in the region, and evaluate the relationships between this sector and the local agri-industries which supply inputs to the fast food chains. In addition, we have sought to identify the global and/or regional linkages which have been generated during the growth periods of the 1980s and 1990s.

Kentucky Fried Chicken in Thailand and China

The first western fast-food chain to be established in Thailand was Pizza Hut, which opened an outlet in Pattaya, the major tourist resort on the southeast coast, in 1980. This was the first Pizza Hut to be established anywhere in Asia, and was selected because of the large number of foreign tourists who visited the resort and who demanded this type of product. But local tastes were quick to adapt and by 1995, there were 85 Pizza Hut outlets in Thailand, with a turnover of US$41.6 million and a net profit of US$4.37 million (*The Nation*, 6 March 1996).

However, it is the KFC chain which is the market leader in Thailand, accounting for 34 percent of fast food sales (*The Nation*, 20 March 1996). The first KFC outlet was opened in 1984, and by 1995, with only 100 stores operating, the chain served over 42 million meals annually, and catered to 100,000 customers each day (*Bangkok Post*, 26 August 1995). By 1997 there were 150 KFC restaurants in Thailand, with a turnover (in 1996) of US$90 million. There are currently three franchise holders of the KFC product in Thailand: the first is Pepsico Restaurants International, which operates over half of the KFC outlets; the second is Thai KFC, a subsidiary of the Central

Group, a major company with particular strengths in department stores and shopping centres. The third is CP KFC Development (Thailand) Ltd, which is a joint venture between Charoen Pokphand (Asia's largest agri-food company and commonly referred to as the CP Group) and the US parent, KFC International, owned by Pepsico Incorporated. Currently, Thai KFC and the CP Group's joint venture share equally in the ownership and operation of the remaining KFC outlets in Thailand (*Asian Franchise and Business Opportunities*, April-June 1997).

A number of the major product lines sold by KFC in Thailand are imported, including french fries and soft drinks. There is a 40 percent duty on french fries which means that compared to the US, food input costs are higher, although this is offset by cheaper labour costs. KFC did attempt to source frozen french fries locally in the mid-1980s, but this appears not to have been successful, largely because of the difficulty in producing the Russet-Burbank variety, which McDonald's has established as the industry standard (*Business in Thailand*, December 1985; *Business Review*, January 1991). However, the most significant input to KFC's operations in Thailand - namely poultry products - are locally sourced. Indeed, the CP Group not only operates the largest number of KFC outlets, but is itself the main supplier of poultry products to these and other outlets under its control. Currently, CP is the largest of the seven major processors of poultry in Thailand and is one of four suppliers of cut-chicken pieces to the KFC chain (*Asian Franchise and Business Opportunities*, April-June 1997).

The CP Group's involvement in poultry production goes back to 1975 when the company initiated its contract farming operations at Sri Ratcha in Chonburi province. The CP Group had long been involved in the production of high quality animal feed, and the establishment of vertically-coordinated poultry production enabled it to both expand the market for its feed and to open up new lines of business in the export of frozen poultry to Japan. The Group's poultry operations laid the groundwork for CP's dominance in agri-food production in the region (for further details see Burch, forthcoming; Goss *et al.*, forthcoming).

However, the Group's involvement in fast food goes further, since the company owns and operates another fried chicken chain in Thailand, Chester's Grill, which had 32 outlets in 1995 and which planned to open a further 15 outlets per year between 1995 and 1998 (*Bangkok Post*, 10 August 1994; 27 December 1994; 4 April 1996). In addition, the CP's ownership of some 800 7-Eleven convenience stores throughout Thailand (which mostly sell food products and which serve nearly a million customers a day), reinforces its substantial capacity in the vertically-coordinated system of poultry production and marketing which it pioneered (*Bangkok Post*, 18 May 1996; *Asiaweek*, 29 August 1997). Thus, the company's impact on the structure and organisation of the Thai poultry industry is clearly substantial (*Business in Thailand*, August 1992; *The Nation*, 12 February 1997).

Turning to China, most of the major North American fast food chains have established operations in order to exploit the huge market opportunities. McDonald's opened its first outlet in Shenzhen province in 1990 and another in Beijing in 1992, at a time when the fast food market was estimated at US$3.5 billion and was reportedly growing at a rate of 20 percent per annum. By 1996, there were 29 McDonald's outlets in Beijing (*Bangkok Post*, 24 April 1992; 23 December 1993; 19 December 1994; 22 February 1995; Yan 1997). McDonald's was followed by Pizza Hut, which had six outlets in 1995. But it was KFC which again pioneered the fast food revolution in China, having opened its first outlet in Beijing in 1987 and its second in Shanghai in 1989. By 1994, the company had 28 outlets and announced plans to establish some 200 restaurants in 45 cities by 1998, involving an investment of US$200 million and the employment of 20,000 workers. By 1997, KFC was on the way to meeting this target, having established 150 outlets in numerous centres (United States Department of Agriculture, 1997). The company's Chinese operations were highly profitable, with outlet sales up to 20 times greater than the average US store. As the president of KFC, John Cranor, stated in 1994 at the opening of the 28th KFC outlet in China (and the 9,000th in the world),

> Asia is KFC's future and nowhere more than in China...By the first decade of the next century China could be our largest market worldwide (*Bangkok Post*, 30 May 1994).

Importantly, in the growth of KFC operations in China, the Thai-based CP Group has again played a significant role, although it is not clear whether the Groups's original investments in poultry production were mainly oriented towards China's domestic market. Undoubtedly, the Group also sought to take advantage of China's lower labour and production costs, and to use it as a production platform for exports of broiler products for Japan (*Bankok Post*, 23 November 1994). In any event, the CP Group began by establishing a company for the local production of animal feed when, in 1979, it set up a joint venture with the US-based Continental Grain to build China's first modern feedmill. Subsequently, the CP Group invested heavily in feedmills on its own account, and by 1995, it was operating 75 such mills (*Bangkok Post,* 7 April 1995; *Business Review,* July 1995).

The CP Group sought to integrate these feedmill operations with the production and marketing of poultry and poultry products for both local and Japanese markets. In the mid-1980s, the Group established three fully-integrated poultry production facilities in the Beijing, Shanghai and Jilin regions, in association with local Chinese companies and in close proximity to its animal feed plants. In the Shanghai region, for example, a joint venture was established between two Songjiang companies and a CP Group subsidiary, Chia Tai. This new company, known as Shanghai Dijiang, operates 50 integrated corporate farms which, by 1995, were producing 50 million chickens per annum, the highest level of output from one operation in China. About 30 percent of output was exported to Japan, and the remainder to markets in Shanghai, Beijing, Zhejiang and Jiangsu. In 1997, Dijiang's exports to Japan were worth US$70 million (*Agrafood Asia*, March, 1998: 8; *Bangkok Post*, 23 November 1994; 7 April 1995; *Business Review*, July 1995; United States Department of Agriculture, 1997).

It was in 1989-90 that the CP Group moved further back into the chain of integration domestically, when it entered into a joint venture with KFC's US parent, Pepsico International, which gave the Group franchise rights over KFC outlets in 13 of China's largest cities. This involved the delivery of 76 million birds per annum, which CP delivered from its local poultry rearing operations. As in Thailand, the CP Group also had franchise rights over the 7-Eleven chain of convenience stores in Shanghai and elsewhere, and established a number of branches of its wholly-owned chain of chicken restaurants known as Chester's Grill. As a result of these and other activities, by 1997, the CP Group accounted for 20 percent of China's chicken market and 10 percent of the animal feed market (*Asian Business*, December 1991; October 1992; *Asian Finance*, 15 November 1988; 15 November 1990; *Business Review*, July 1995; United States Department of Agriculture, 1997).

While per capita consumption of chicken meat in China remains low at 4.3 kilograms per annum (compared to 9.2 kilograms in Thailand, 13.7 in Japan and 46.6 in Hong Kong), this only serves to demonstrate the perceived scope for further growth and the potential for even greater returns to the CP Group and the fast food chains with which it is linked (*Bangkok Post*, 23 November 1994; International Finance Corporation, 1995). But this, of course, depends upon continued growth in incomes and output which, in the wake of the 1997 crisis, is now questionable.

McDonald's in Malaysia and Thailand

As in Thailand and China, it was the KFC chain which pioneered the fast food sector in Malaysia, opening its first outlet in 1973, and it remains the market leader. The KFC chain has 190 outlets and accounts for 58 percent of the country's fast food market, while McDonald's with 100 outlets has 20 percent. KFC in Malaysia also owns the Pizza Hut franchise, which accounts for a further seven percent of the fast food market (*Malysian Business*, 1 June 1997; *The Nation*, 13 August 1996). Part of KFC's success is due to the fact that chicken is universally accepted by the three main communities in Malaysia (Malays, Chinese and Indians), whereas about 10 percent of the population (those of Chinese or Indian descent who are Taoist or Hindu) abstain from beef consumption. At the same time, McDonald's is somewhat disadvantaged by the fact that the dominant Malay population is largely Muslim and therefore shuns the pork products which elsewhere form part of the chain's breakfast menu.

McDonald's opened its first outlet in Malaysia in 1982, but did not start to show a profit until 1990. The franchise rights are held by Golden Arches Restaurants, which is jointly owned by the McDonald's Corporation (a 49 percent share), Mohamed Shah Abdul Kadir (McDonald's managing director, who holds a 25 percent share), and Tan Sri Vincent Tan Chee Yioun (chief executive officer of the Berjaya Group conglomerate, with a 26 percent share).

With the exception of its kitchen equipment and some food products, McDonald's in Malaysia has long sourced most of its inputs from local suppliers, although these are not necessarily Malay-owned companies. Thus, the local supplier of fish, beef and some chicken products to the Malaysian outlets is MacFood Services, which is a subsidiary of the US-based Keystone Food Corporation, while some of McDonald's chicken products are sourced from CP Poultry, a local subsidiary of the CP Group. McDonald's has also established a project for the local production of lettuce, in association with the Malaysian Agriculture Research and Development Institute (*Malaysian Business*, 1-15 January, 1992).

In fact (as Table 6.3 shows), a large number of Malayasian-based companies emerged as significant suppliers of inputs to McDonald's stores to the wider region. Golden Arches Restaurants pursued a conscious policy of expanding the production in Malaysia of a wide range of products required by McDonald's outlets throughout the Southeast Asia region. This policy of

Table 6.3   Malaysian-based suppliers of McDonald's products, 1996

Supplier	Product	Export countries
MacFood Services	Beef, fish, chicken	Singapore, Hong Kong
Havi Food Services	Buns	Indonesia
Campbell Soup Southeast Asia	Condiments (chilli, sauce)	Singapore, Taiwan, Hong Kong, Thailand, Taiwan
Tien Wah Press	Packaging for bone-in chicken and nuggetts	Indonesia, South Korea, Singapore, Hong Kong, Thailand, Taiwan
Sekoplas Industries	Plastic bags	Singapore, Philippines
MarkmasPak-Print	Drink carriers	South Korea, Singapore
Hong Kong Printing	Fries, hash brown, apple cartons, 2 or 4 ring-type drink holders	Thailand, Hong Kong, South Korea, Indonesia, Singapore, Taiwan, New Zealand, Australia
Scott Paper	Napkins and junior roll	Hong Kong, Singapore, Indonesia, Taiwan, South Korea, Japan
Guppy Plastic Industries	Plastic cutlery kit, recycled and virgin in-store trays	New Zealand, Taiwan, Hong Kong, Singapore, South Korea, Japan, Indonesia
Creative Packaging	Hash brown bag	Singapore, Taiwan

Source: Malaysian Business, 1 June 1997

regional sourcing was adopted in order to maximise the local production of inputs, thereby assisting the Malay company overcome its initial losses and to move into profit by reducing its unit costs (*Malaysian Business*, 1-15 January 1992: 15). In 1996, Malaysian and Malaysian-based companies were supplying a large number of McDonald's stores throughout East and Southeast Asia and Australasia with the standard inputs they required for their operations.

In the case of Thailand, McDonald's is the second largest of the fast food chains, with 53 outlets in 1997. From its inception in 1984, the company owning the McDonald's franchise in Thailand - the McThai Company - has shown a profit. It has also long sourced some 80 percent of its inputs from local suppliers. Initially, the hamburger patties were supplied by an Australian company, and some packaging products for chicken and other food products are sourced from Malaysian suppliers (see Table 6.3). However, because of the high rates of duty.on many imported food products, the company has sought, where possible, to source from local suppliers (*Bangkok Post,* 13 March 1997; *Business in Thailand,* December 1985; *Business Review,* 1 August 1988).

As with the Malaysian operation, the McDonald's company has also sought to maximise output and economies of scale by utilising local Thai suppliers in the production of commodities for company outlets in the wider region. In 1994, for example, United Flour Mill Company (UFM), a division of the large Metro Group conglomerate and one of the largest wheat flour manufacturers in Thailand, was contracted as the supplier of dessert pies to all McDonald's restaurants in the Asia-Pacific region. Previously, these products were supplied by the US-based Bama Pie Company, but import duties added to costs and the long distances involved in transportation resulted in a loss of freshness. Upon appointment as a supplier to the Asia-Pacific region, UFM established new productive capacity which enabled it to increase its production five-fold, to 75 million pies per year (*Bangkok Post,* 28 April 1994).

However, there seem to have been problems with the local sourcing products of french fries. The McThai company did attempt to do this soon after it was established in 1984, when a local trading firm, Bara-Windsor, established a joint operation with the Idaho-based supplier of french fries to McDonald's in the US, J. R. Simplot. Bara-Windsor imported Russet Burbank seed from J. R. Simplot and supplied it to farmers in Northern Thailand. There were 200 contract growers organised over 65 hectares, who in 1985 produced 1,000 tonnes of potato. However, the new and complex techniques of production and crop management, combined with the large scale volumes, high standards and quality control requirements stipulated by McDonald's, seemed to impose conditions which were not suited to small-scale cultivation, and local production ceased.

Local production of french fries was a problem not just in Thailand, but throughout the region. As a consequence, the US is now a major exporter of frozen french fries to a large part of the fast food industry throughout East and Southeast Asia. Japan, South Korea, Hong Kong, Taiwan, the Philippines, Singapore and Malaysia currently rank among the largest markets for this product. In 1992, for example, these seven Asian countries imported US frozen french fries valued at US$105 million,which represented some 88 per cent of total US exports of this commodity, and by 1995 Japan alone imported US$98 million of frozen french fries from the US (*The Australian*, 4 July 1985; *Business in Thailand*, May 1985; May 1996).

The Asian currency crisis and the future of Fordist food

The crisis which hit the economies of a number of Asian countries in 1997 brought to an end the 'miracle' of rapid economic growth which had been sustained since the 1980s. Countries which had for many years operated on the assumption that such growth could continue unabated suddenly found themselves confronting the prospect of negative growth, reduced investment and growing unemployment. In Indonesia, the worst affected of the Asian economies, it was expected that the economy would contract by some 10 percent in 1998, with inflation running at 80 percent by the end of the year (*Sydney Morning Herald*, 26 June 1998). It was further predicted that per capita income would fall from US$1200 to US$750 and that the number of absolute poor would increase from 24 million to 42 million (*Far Eastern Economic Review*, 25 December 1997; 1 January 1998; 19 February 1998). In Thailand, it was expected that that GDP would contract by three to five percent in 1998, and that real incomes would decline as inflation increased to 12 percent per annum while wages would only increase by six percent (*The Australian,* 20 April 1998; *Bangkok Post*, 3 November 1997; *Far Eastern Economic Review*, 19 February 1998). In the opinion of some influential observers, such as the World Bank, it was likely that these early projections might need to be revised as a second wave of the crisis developed and further undermined already fragile regimes (*The Australian*, 16 June 1998).

These conditions have had a major impact on the profitability of the fast food companies throughout the region, as well as the agri-food sectors supporting them. Before the crisis, the fast food sector in most of the countries of the region was very dynamic, and had demonstrated growth in the order of

25 to 30 percent per annum up to mid-1997. But by November 1997, most of the chains in Thailand had begun to feel the impacts of falling incomes and growing unemployment. In January 1998, one of McDonald's leading outlets in Bangkok, located on the once-booming retail centre around Thanon Asoke, reported a 20 percent decline in profits, while KFC reported a 10 percent drop in sales. All the leading chains resorted to significant price reductions in order to generate some income in the face of falling demand. The problems of McDonald's outlets in the region became so acute that the US parent temporarily waived fees to franchisers (*Bangkok Post,* 15 September 1997; 3 November 1997; 26 January 1998; 27 March 1998).

Under these circumstances, local and off-shore agri-food companies supplying inputs to the fast food chains have also come to experience difficult trading conditions. This is especially evident where the declining value of Asian currencies has led to significant increases in the price of imported products. In Indonesia, McDonald's (which had 38 outlets in 1995) ceased to sell french fries (a fully imported food item) and menu prices were reduced by half (*Sydney Morning Herald*, 14 February 1998). Basic local food prices rose by 75 percent in the space of six months, and by early 1998, the Indonesian poultry industry, a key supplier to the country's fast food outlets, was operating at 30 percent capacity. Much of Indonesia's animal feed was imported, and with feed costs having doubled in the space of only a few months, the costs of production exceeded the selling price of live birds (*Agrafood Asia*, April 1998). The picture was much the same elsewhere. Between March and May 1998, for example, 30 percent of Malaysia's poultry farms had shut down, following an 80 percent increase in the price of imported animal feed, leading to rising prices and falling demand (*Agrafood Asia,* May 1998).

Such developments raise questions about the continued viability of the fast food sector in the medium term, and about the prospects for growth in the future. Prior to the 1997 crisis, all the major chains had significant plans for continued expansion. McDonald's in Thailand planned to double the number of its stores by 2000 (*Bangkok Post*, 13 March 1997), while in Malaysia, the company envisaged a four-fold increase in the number of outlets, taking it to 400 restaurants by the year 2000. In China there was a plan to open a total of 600 McDonald's outlets by the same date (*Malaysian Business*, 1 June 1997: 24; Yan 1997). Also in China, Pizza Hut was planning to open 300 new restaurants between 1995 and 2005. The fast food industry's public response to the crisis has been to reaffirm its plans for expansion. McDonald's reportedly still intended to open six new restaurants in Bangkok in 1998, and also announced its plan to increase the number of outlets in Japan to 10,000 by the year 2006, up from the current 2,500. Similarly, in November 1997, Pizza Hut reiterated its intention to open 50 new outlets in Thailand in 1998 (*Agrafood Asia*, June 1998; *Bangkok Post*, 3 November 1997; 27 March 1998). However, it remains to be seen whether these bold plans will eventuate.

Specificities and contradictions

While the impacts of the crisis on fast food companies within the region raise significant concerns, it is important to acknowledge that the impacts may be highly variable insofar as the crisis affects some countries and/or local franchisers more than others. For example, where there is a significant level of imported inputs going into the production of fast food products (e.g. animal feed), then the local impacts will be extremely significant. On the other hand, in the case of Malaysia, McDonald's policy of encouraging the growth of local suppliers who are also able to serve the needs of McDonald's outlets throughout the Asia-Pacific region, may cushion it from the worst effects of the meltdown. McDonald's in Malaysia will be able to source much of its needs locally from a group of suppliers who are able to provide the benefits of scale, rather than importing these items with a devalued currency. Conversely, however, those McDonald's outlets in the region which are sourcing from Malaysia are having to pay more for their imports at a time when falling incomes are likely to lead to reduced demand. As a consequence, these companies will either reduce their demands on the Malaysian suppliers or seek cheaper sources elsewhere (an unlikely outcome in the short term). In any event, whatever course of action is followed, it is likely to increase unit costs of production in Malaysia, and in that way, impact also on the apparently well-insulated operations of McDonald's in that country. Again, some Thai-based suppliers (e.g. of poultry products) may even benefit from the crisis to some extent, because the cutbacks in poultry production (combined with the devaluation of the baht) have led to significant price increases which Thai producers are well-positioned to exploit. Indeed, the US parent of McDonald's has negotiated with the Thai supplier of meat products, McKey Food Service, to increase five-fold the supply of chicken products to McDonald's outlets in Hong Kong and Singapore (*Bangkok Post*, 27 March 1998).

But of course, while some fast food chains made extensive use of local inputs, the policy of regional sourcing combined with the need to import certain products which could not be produced locally, meant that to varying degrees, *all* chains were dependent upon the ability to source some food and raw materials from other countries inside and outside the region. This tendency was further encouraged through the stabilising mechanism of exchange rate controls undertaken by various central banks in the name of economic 'prudence'. This allowed the corporate community in Southeast Asia relatively easy access to overseas capital, thereby discouraging import substitution practices and relying instead on the comparative advantage of cheap labour and resources in line with the NIC 'model' of development (World Bank, 1993). In this sense, these policies were a departure from the classical notions of Fordism as the national economy was not regulated to the same degree (Boyer, 1997: 21-22).

However, these forms of state intervention did encourage certain industrial developments through relatively cheap technology transfer and foreign investment, thus enabling the growth of manufacturing and service industries (e.g. textiles, electronic components and tourism). These processes stimulated consumer demand as industrial employment increased, but for products produced outside the region. In terms of food, for example, this was represented by increasing imports of dairy products and frozen french fries as the standardised food and quality requirements of the fast food industry were unable to be met through local production. While Malaysian paper producers with access to low cost raw materials could produce cheap packaging and export it to the region, the demands of high quality food products for the 'regimes' imposed within the various fast food industries could not be met as easily in the domestic production sphere. As a result, Australian, US and European producers of beef and dairy commodities enjoyed extremely high levels of growth in the demand for their products. In addition, the growth of live cattle exports generated further significant profits for Australian producers (*Sydney Morning Herald*, 15 November 1996).

Conclusion

What, if anything, do such developments tell us about the nature of 'peripheral Fordism'? More importantly perhaps, what are the implications for the broader conception of the NIC model of development, and its ability to be sustained over time? Is the current crisis in the Southeast Asian region merely a perturbation or a 'correction', which will be followed by a recovery and continued growth of the kind experienced over the 1980s and early 1990s? Or is there some more fundamental transformation occurring which signals the end the NIC model and the transition to some new situation?

In terms of the fast food systems of Southeast Asia, the 1980s and 1990s witnessed the growth of a 'peripheral' Fordism similar to - but at the same time different from - the experience of Fordism within North America and Europe. One of the most important distinguishing features of the fast food sector in this 'peripheral' Fordist model was that McDonald's hamburgers and KFC chickenburgers were not items of everyday consumption in the same way that they were in the US or Australia. These products might have been routinely eaten by wealthier middle-class patrons, but were an occasional and expensive treat for lower-income groups in many countries. As Yan (1997: 54) notes, in discussing McDonald's in China,

> eating at McDonald's is still a big treat for low-income people, and...as of 1994, a dinner at McDonald's for a family of three normally cost one-sixth of a worker's monthly salary.

In Thailand in 1997, a Big Mac represented almost one-third of the daily minimum wage, while prices at KFC were about the same; the average meal at KFC was 40 to 45 baht per person at a time when the minimum daily wage was pegged at 153 baht, indicating that few of the many workers on the basic wage could ever afford to consume this product (*Bangkok Post*, 13 March 1997; 3 November 1997). In one sense then, what counts as Fordist food in one context is not necessarily Fordist food in another. While the inputs and production processes may be exactly the same in both industrialised and less-developed countries, the hamburger in China is not Fordist in the sense that it is a cheap, widely-consumed commodity which is the product of a social system in which the producers of mass-produced goods are also the consumers of those goods.

This point is related to another distinguishing characteristic which sets 'peripheral' Fordism apart from its western counterpart. This is the fact that in the NICs of Southeast and East Asia, it is difficult to equate the relationship between government, industry and the population as the same kind of corporatist accomodation or *rapprochement* that was characteristic of Fordism in the liberal capitalist democracies. Keynesian intervention by elected governments in a long period of post-war reconstruction and economic growth represented a different social order from that prevailing in most NICs, where authoritarian governments, often backed by the military and engaged in the suppression of local labour movements, were concerned to industrialise rapidly on the basis of labour-intensive industries, where it was important to 'get the prices right', i.e. maintain low wages and a competitive edge. In this model, the Fordist principal of paying increasing wages so that workers are able to consume the commodities they produce has little relevance, especially in an export-oriented economy in which consumers are to be found overseas rather than at home.

There is little doubt that over the 1980s and early 1990s, incomes and purchasing power in most of the NICs increased and the incidence of poverty declined, but the distribution of income often became more unequal. In Thailand, for example, the share of income accruing to the bottom 20 percent of the population fell from 4.6 percent in 1988 to 4.5 percent in 1996. In contrast, the share of the top 20 percent increased from 54.2 percent in 1988 to 55.4 percent in 1996 (Kakwani and Krongkaew, 1996). In addition, a high incidence of absolute poverty usually persisted in the more distant and isolated regions which remained dependent on agriculture. But the most important point at issue here is the extent to which the recent currency crisis has wiped out the relativelty modest gains by the urban and rural poor in recent years, and increased the incidence of absolute poverty.

This general picture suggests that the manifestation of some Fordist characteristics within the political economy of the less developed countries of East and Southeast Asia, have to be treated with a degree of caution when considering the implications for theory and practice. The emergence of fast food chains in the region, for example, may indicate some kind of transition by these countries towards a different social order which is akin to that experienced by the industrialised countries in the post-war period. But such an interpretation would, in our view, be quite misleading. The growth of fast food in Southeast Asia represents a different experience, one conditioned by the patterns of growth and poverty reduction which were not only varied, but which may have come to an end.

Certainly, the events of 1997 mean that, for now, the dream of a continuing expansion of mass consumption in Asia is over. The other issue, which is little discussed but which is clearly in the minds of regional analysts, concerns the poor - the bulk of the region's population whose transference to wage labour systems was premised upon an expanding base of waged-labour opportunities. The advance of fast food in Southeast Asia, and the productive and consumption orders associated with it, were moments in a specific historical period which can neither be repeated through policy prescriptions nor re-engendered through a 'recovery'. This being the case, it opens up the prospect that the promise and expectation of increasing affluence will never extend beyond yesterday's hamburger, and the social unrest that is already beginning to manifest itself in the region can only grow. For Southeast Asia, 1997 could be as significant a transformation as 1989 was for Eastern Europe.

Notes

1   We are grateful to the participants of the Akaroa meeting of the Agri-food Research Network, whose comments helped in refining our arguments. Needless to say, all errors are our responsibility. We also acknowledge partial funding from an Australian Research Council Large Grant. Finally, Jasper Goss would like to thank the Sir Robert Menzies Centre for Australian Studies, Institute of Commonwealth Studies, London, and the Australian Bicentennial Scholarship Scheme for support that contributed to an earlier version of this chapter.

2   Not all food production resembles this process and certainly the debates initiated by Goodman *et al.* (1987) over the specificities of food remain relevant. The biological conditions of food production are clearly an important process, although the level of priority to be accorded to this insight remains contested (see the debates between Ben Fine and Michael Watts for further analysis of this issue: Fine, 1994a; Fine, 1994b; Watts, 1994). For the purposes of our argument, it is perhaps best to acknowledge the variations of importance that the biological conditions of food play within different sectors of production in the industry. In this way, we can view the homogenisation of particular food sectors (e.g. fast food) as indicative of greater controls over the biological process which inversely relate to the significance of the biological process - i.e. the greater the homogenisation, the less biological variance. Of course, the result of this process is that greater resources must be utilised to decrease the influence of 'biology'.

3   The NICs included in the calculation of these figures are Hong Kong, Indonesia, Malaysia, Singapore, Thailand and South Korea. For the period between 1985 and 1993, Taiwan averaged 8.1 percent annual GDP growth (Chu, 1996: 205).

References

Agrafood Asia, No. 47, March 1998; No. 48, April 1998; No. 49, May 1998; No. 50, June 1998.

Aglietta, M. (1979), *A Theory of Capitalist Regulation: The US Experience*, New Left Books: London.

Amin, A. (1994), 'Post-Fordism: Models, Fantasies and Phantoms of Transition', in Amin, A. (ed.), *Post-Fordism: A Reader*, Blackwell: Oxford, pp. 1-39.

Arrighi, G. (1994), *The Long Twentieth Century*, Verso: London.

Asian Franchise and Business Opportunities (Singapore), January 1996.

Asian Business (Hong Kong), December 1991.

The Australian (Sydney), 4 July 1985; 15 October 1997; 14 April 1998; 16 June 1998.

Bangkok Post (Bangkok), 10 August 1994; 27 December 1994; 15 May 1995; 26 August 1995; 4 April 1996; 11 March 1997; 13 March 1997; 1 September 1997.

Barnett, R. and Cavanagh, J. (1994), *Global Dreams*, Simon and Schuster: New York.

Boyer, R. (1997), 'How Does a New Production System Emerge?', in Boyer, R. and Durand, J.-P., *After Fordism*, MacMillan: London, pp. 3-63.

Burch, D. (forthcoming), *Transnational Agribusiness and Thai Agriculture: Globalisation and the Political Economy of Agri-food Restructuring*, a report to the Institute of Southeast Asian Studies: Singapore.

Business in Thailand (Bangkok), August 1992.

Business Review (Bangkok), January 1991; July 1995.

Courier-Mail (Brisbane), 19 July 1997.

Chu, J. (1996), 'Taiwan: A Fragmented Middle Class in the Making', in Robison, R. and Goodman, D. (eds), *The New Rich in Asia: Mobile Phones, McDonald's and Middle-Class Revolution*, Routledge: London, pp. 205-222.

Far Eastern Economic Review (Hong Kong), 25 December 1997; 19 February 1998.

Fine, B. (1994a), 'Towards a Political Economy of Food', *Review of International Political Economy*, Vol. 1, No. 3, Autumn, pp. 519-545.

Fine, B. (1994b), 'A Response to My Critics', *Review of International Political Economy*, Vol. 1, No. 3, Autumn, pp. 579-586.

Fine, B., Heasman, M. and Wright, J. (1996), *Consumption in the Age of Affluence*, Routledge: London.

Goodman, D., Sorj, B. and Wilkinson, J. (1987), *From Farming to Biotechnology*, Blackwell: Oxford.

Goodman, D. and Watts, M. (1994), 'Reconfiguring the Rural or Fording the Divide? Capitalist Restructuring and the Global Agro-Food System', *Journal of Peasant Studies*, Vol. 22, No. 1, October, pp. 1-49.

Goss, J., Burch, D. and Rickson, R. (forthcoming), 'Agri-Food Restructuring and Third World Transnationals: Thailand, the CP Group and the Global Shrimp Industry', *World Development*.

Gramsci, A. (1971), *Selections from the Prison Notebooks*, International Publishers: New York.

Harvey, D. (1990), *The Condition of Postmodernity*, Blackwell: Oxford.

Hoogvelt, A. (1997), *Globalisation and the Postcolonial World*, Macmillan: London.

International Finance Corporation (1995), *The World Poultry Industry*, World Bank: Washington D.C.

Jessop, B. (1990), *State Theory: Putting Capitalist States in their Place*, Pennsylvania State University Press: University Park PA.

Kakwani, N. and Krongkaew, M. (1996), 'Income Distribution', *Bangkok Post 1996 Economic Review*, December.

Lash, S. and Urry, J. (1987), *The End of Organised Capitalism*, Polity Press: Cambridge.

Lawrence, G. and Vanclay, F. (1994), 'Agricultural Change in the Semiperiphery: The Murray-Darling Basin, Australia', in McMichael, P. (ed.), *The Global Restructuring of Agro-Food Systems*, Cornell University Press: Ithaca NY, pp. 76-103.

Lipietz, A. (1982), 'Towards Global Fordism?', *New Left Review*, No. 132, pp. 33-47.

Lipietz, A. (1987), *Mirages and Miracles: The Crises of Global Fordism*, Verso: London.

Love, C. (1987), *Beneath the Golden Arches*, Bantam Books: Toronto.

Malaysian Business (Kuala Lumpur), 1-15 January 1992; 1 June 1997.

Matthews, J. (1989), *Tools of Change: New Technology and the Democratisation of Work*, Pluto Press: Sydney.

The Nation (Bangkok), 20 March 1996; 9 August 1996; 13 August 1996; 12 February 1997.

Sydney Morning Herald, 15 November 1996; 14 February 1998.

Teeple, G. (1995), *Globalisation and the Decline of Social Reform*, Garamond Press: Toronto.

United States Department of Agriculture (USDA), *Agricultural Trade Reports: China - Annual Poultry Report*: Washington D. C.

Watson, J. (1997), 'Introduction: Transnationalism, Localisation, and Fast Foods in East Asia', in Watson, J. (ed.), *Golden Arches East: McDonald's in East Asia*, Stanford University Press: Stanford, pp. 1-38.

Watts, M. (1994), 'What Difference Does Difference Make?', *Review of International Political Economy*, Vol. 1, No. 3, Autumn, pp. 563-570.

World Bank (1993), *The East Asian Miracle: Economic Growth and Public Policy*, Oxford University Press: New York.

World Bank (1997), *World Development Report 1997: The State in a Changing World*, Oxford University Press: New York.

Yan, Yungxiang (1996), 'McDonald's in Beijing: The Localization of Americana', in Watson, J. (ed.), *Golden Arches East: McDonald's in East Asia*, Stanford University Press: Stanford, pp. 39-76.

PART II
RESTRUCTURING WITHIN AUSTRALIAN AND NEW ZEALAND AGRI-FOOD SYSTEMS

7 Capitalism, the state and *Kai Moana*: Mäori, the New Zealand fishing industry and restructuring

KHYLA RUSSELL AND HUGH CAMPBELL

Introduction

It is testimony to the hegemony of pastoral farming in the economic and ideological history of New Zealand that so little has been written about New Zealand's fishing industry and even less about the relationship between Mäori and fishing.[1] Yet, while pastoral farming was central to the control of land and the consequent disenfranchisement of Mäori throughout the colonial history of New Zealand, the control over fisheries resources has also been crucial and deserves consideration. In fact, the fishing industry in New Zealand can be analysed in terms of five distinct historical phases which render considerable insight into both the internal politics of resource control in New Zealand and the changing political and economic relationship between New Zealand and the rest of the world.

There are aspects of New Zealand's fishing industry which are unusual in comparison to the historical analysis of emergent food commodity systems in other settler states. As the original inhabitants of New Zealand, and the most adept and established fishers of New Zealand's water, Mäori played a very important role in the early colonial history of fishing. Then, rather than fading from view as so many indigenous peoples did throughout the settler states, Mäori re-emerged to contest the developing export fishing industry in New Zealand post-1970. This re-emergence and contestation has become the battleground on which numerous political and economic struggles are being played out and indicate that New Zealand's indigenous people still have a major stake in the ongoing history of one key food commodity chain.

Contact and the colonial economy (1835-1866)

Prior to colonisation, fisheries and fishing grounds in New Zealand, like other aquatic resources, were not common property available to all. A complex set of rights determined who could harvest certain fish species, fish in particular areas and at what times harvesting could occur (Te Ohu Kai Moana, 1997: 2).[2] These rights were organised around Mäori kinship groupings. *Whänau*[3] held fisheries' rights over small eel weirs on streams, small *waka* (canoes), fishing grounds and shellfish beds beside their *kaik*.[4] *Hapü*[5] exercised control over large eel weirs on main rivers, larger *waka* and sea faring vessels and specified fishing grounds. *Iwi*, the greatest social grouping, incorporated the rights of its composite kin groups, the property of which included various *hapü* lands, lakes, rivers, swamps and streams within them, beaches, mudflats, rocks, reefs, off-shore islands and ocean. All major fishing expeditions (as with other food planting and harvesting) were undertaken at the *iwi* level (Memon and Cullen, 1992: 155; Taiapa, 1996: 13-24; TOKM, 1997: 2). The control of human behaviour in relation to fisheries and their sustainability was governed by *tapu* (restriction), *makutu* (ritual forces) and *rähui* (prohibition against over use or trespass), all of which were applied by *Rakatira* (chiefs). Such mechanisms governed when, and by whom, fish were harvested. *Tapu* and *makutu* protected fisheries through constraint and extent of use, while *rähui* prevented fish being taken out of season. Such measures as instituted by Mäori can be argued to be a sustainable use of a renewable resource.

A key characteristic of fishing tribes was their dependence upon fishing, as fisheries had subsistence, commercial, and cultural importance. All of these were paramount to individual and community survival, as well being an important component of socio-economic networks between *hapü* and *iwi*. For Mäori, fishing was (and to a large extent still is) a dynamic expression of *rakatirataka*, which involved complex knowledge of geography, oceanography, the biology of fish species and their food chain/supplies (Beattie, 1994: 138). Such knowledge included harvesting and processing technology, regulation and management of harvests and habitats, as well as ethical and spiritual practices and values (Beattie, 1994: 138; Tau, 1995: 14).

O'Regan in TOKM (1997: 2) states that pre-contact Mäori fishing activities existed in a form that was 'capable of adaptation to commercial uses in Western terms' - particularly as indicated by inter-tribal trade which easily expanded into trade with early European arrivals. This process of commercialisation of existing Mäori structures culminated in direct trade, both nationally and internationally, for quite some time after the signing of the Treaty of Waitangi in 1840. Later actions by the Crown in New Zealand clearly overlooked the extent and value of commercial Mäori fishing activities during this first period of contact - activities which were crucial for the sustenance of newly arrived Päkehä colonists.

It can be argued that the Mäori commercial trade with Päkehä settlers was based on an industry that had developed indigenously and did not rely on borrowed technology or knowledge from settlers. This claim is supported in two important fora. The Waitangi Tribunal (Waitangi Tribunal, 1988: 196-7) cited recent historical and archaeological research, as well as oral findings, demonstrating that there was a substantial indigenous fishing industry in existence prior to 1840 (Memon and Cullen, 1996: 255). The New Zealand Law Commission (NZLC) (1989: 27-30) found the same, citing six historical and anthropological Päkehä accounts to back up their claim supporting the widespread existence of inshore and inland fishing by Mäori. A further three sources are cited, as well as the Muriwhenua report, for evidence of off-shore fishing activities by Mäori on both islands (Waitangi Tribunal, 1988: 30-32, 152, 196-7). The NZLC (1989: 27) went on to assert:

> On the testimony of European explorers from Cook to du Fresne, Mäori fishing equipment and methods in the early nineteenth century were in many respects superior to those of Europe. Numerous writings testify to the special place of fisheries in [the] Mäori economy and culture in pre-European times, throughout the nineteenth century and even later.

Memon and Cullen (1996: 255) contend that early Päkehä arrivals noted that sizeable coastal Mäori communities whose economy was based on fishing existed in several areas. It is clear that at the time of early colonial activity there was an important commercial supply of fish to settlers by Mäori. Furthermore, an international trade in fish also took place as, according to Kai Tahu knowledge, Tuhawaiki exported both flax and dried fish to Sydney prior to 1840 and for some years after the signing of the Treaty of Waitangi. The veracity of this little-recognised historical situation is also supported by the NZLC (1989).

Marginalisation and expropriation of commercial Mäori fishing (1866-1930s)

The commercial trade in fish by Mäori appears initially to have had the tacit approval of the Crown. However, both Walker (1992: 120) and Orange (1987: 263) contend that this commercial position was slowly eroded by successive government legislators. Despite Treaty guarantees in Article II of *undisturbed possession of (lands, forests, estates and) fisheries* (Orange, 1987: 263; Walker, 1992: 120), these legislators subscribed to the view that Mäori rights were for non-commercial activities only (TOKM, 1997: 4; Memon and Cullen, 1996: 254). Consequently, Walker (1992: 103) argues that the appropriation of Mäori land was equalled by 'expropriation of their fisheries'.

The first such inroad was the Oyster Fisheries Act 1866, which granted leases over oyster beds in Manukau exclusively to non-Mäori for commercial purposes. This action was legitimised by the claim by government that such actions were necessary to prevent resource depletion - alleging Mäori over use. The Act was extended to the extensive Fouveaux Strait oyster beds in 1869 (Walker, 1992: 103-104) after the discovery of these beds by Kai Tahu in conjunction with Charlie Britt (Beattie, 1994: 158). Once again, the 1869 Amendment granted exclusive commercial licences to non-Mäori despite the role of Kai Tahu in the discovery of this resource (Walker, 1992: 103). The then Northern Mäori MP protested in parliament at the granting of such licences over what had been customarily harvested oysters. A clause was consequently inserted within the Act purportedly as a safeguard for Mäori rights, but which declared Mäori oyster reserves could *only* exist where they were near village or Pä sites. It was also stipulated that the Mäori gathering rights were for purposes of consumption only, forbidding sales of shellfish from these reserves (Waitangi Tribunal, 1988: 81). This Act, and others which followed, effectively established the principle that Mäori fishing rights were confined to subsistence activities and this ensured that Mäori would not be able to commercially compete with Päkehä in the fishing industry (Kelsey, 1993: 260).

Once salmon and trout were introduced, the Crown assumed control over river and lake fisheries. The 1867 Salmon and Trout Act provided Crown protection for these species (Waitangi Tribunal, 1988: 83). Walker (1992: 104) argues the protective element was illusory, while the control element was 'absolute', in regard to both inland and marine fisheries. Introduction of exotic species within lakes and rivers saw them consume many existing indigenous species, so that Mäori began turn to introduced species for subsistence purposes. The result was a legal contest between Mäori and the Crown which resulted in a law which directly counteracted the rights granted under Article II of the Treaty of Waitangi. Initially, legal action was taken against Mäori as they were catching introduced fish with no licences. The Mäori response was that Treaty guarantees still applied and therefore legal action dishonoured Article II. Consequently, the ordinary court referred the matter to the then Native Land Court which decided in favour of customary rights, as it applied to inland fisheries (Waitangi Tribunal, 1988: 85). Finally, to counteract this, the government passed the Harbours Act 1878, in which section 8 directly invalidated customary fishing rights - thereby contravening Article II of the Treaty of Waitangi (Waitangi Tribunal, 1988: 85).

This action relating to salmon and trout set a precedent for the revocation of Mäori rights granted under Article II of the Treaty. Mäori petitioned parliament in response to the constant erosion of fishing rights back to subsistence levels - principally through State legislation - and the emergence of Päkehä capital interests in fishing resources still protected by customary fishing rights (Walker, 1992: 104). Complaints by Mäori for redress 'in respect of fisheries and fishing grounds were frequent, forceful and often futile' (NZLC, 1989: 9). They submitted a claim of ownership over certain lakes to the Stout-Ngata Land Settlement Commission of 1907, claiming customary rights to fresh-water fish for subsistence and arguing that exotic species 'derogated former ones' (Walker, 1992: 104). The Court of Appeal eventually referred this claim to the Native Land Court which delayed the hearing until 1920. The Crown then extinguished customary title in exchange for forty trout licences and a six thousand pound annuity. As with the oysters from the reserves, no selling of trout was permitted by Mäori, though the right of tribes to take indigenous fish species for domestic consumption was allowed (*Te Wänanga*, 1929: 137-8). Walker (1992: 105) cites similar cases where Mäori were unsuccessful in salvaging something of their Article II Treaty guarantees in relation to both customary rights or precluding them from undertaking commercial fishing activities.

Colonisation fragmented Mäori communal ownership of land and resources, replacing it with individual ownership and an economy geared to commodity production for export. Simultaneously, assimilationist policies were pursued that ultimately marginalised Mäori both economically and socially (Memon and Cullen, 1992: 158). After the 1860s land confiscations, the Mäori agricultural and fishing economy rapidly declined (Hargreaves, 1960: 355). Despite this, New Zealand continued to have what is described as a dually structured fishing industry, due to differential participation of Mäori and Päkehä fishers. Memon and Cullen (1996: 256-258) argue that Päkehä hegemony over fishing increased through full-time commercial fishing facilitated by access to capital which allowed the use of superior technology. At the same time, those Mäori fishers who were still engaged in commercial fishing became progressively marginalised in a similar way to the marginalisation of Mäori farmers (Memon and Cullen, 1992: 258). The result was a state-legitimated hegemony of Päkehä over fishing and farming, and the marginalisation of Mäori commercial fishing to the margins of the industry. The position of Mäori as the original exporters of fish from New Zealand was a distant memory.

The New Zealand government began to actively manage fisheries with the 1908 Fisheries Act, which remained in force, with various amendments, until 1983 (Clark *et al.*, 1988: 325). According to Le Heron and Pawson, this Act was considered necessary because the industry was purported to be in a continuous 'state of crisis for most of its European history'. There were two main reasons for this crisis: the uncertainty of its commercially sustainable life; and the future of fisheries resources in relation to Waitangi Tribunal claims, particularly as they applied to the status of fisheries resources as tradeable property rights (Le Heron and Pawson, 1996: 154). While the second of these two 'crises' will be examined in detail later, the response of the government to concerns over the long term sustainability of commercial fishing was to establish a range of legislative controls. From 1938 to 1963, control of the country's in-shore fisheries was managed through a system of restrictive licensing which set the number of vessels able to operate, established extensive controls over gear type, and designated the areas in which licensed fishers could operate (Cullen and Memon, 1990: 50).

Re-commercialisation of Mäori fishing (1940s-60s)

A number of writers have argued that commercial exploitation of New Zealand's marine fishery was relatively benign prior to the 1960s, as the industry was very small (Memon and Cullen, 1992: 256; Walker, 1992: 106, 326). In 1964 however, the government established a Fishing Industry Board with the aim of promoting the industry's expansion, and adding to New Zealand's gross national product (GNP) (Cullen and Memon, 1990: 49; Walker, 1992: 106). This was achieved by removing restrictions on the number of vessels allowed to operate. By the start of the following decade (1970s) foreign vessels began exploiting New Zealand's deep-water and inshore fisheries (Clark *et al.*, 1988: 325; Cullen and Memon, 1990: 46). During the time of licensed control of inshore fisheries, the state was unable to exert any control over off-shore fishing due to the existence of the 19-kilometre limit of sovereignty over the oceans. Nevertheless, the government still encouraged the expansion of the fishing industry investment through 'investment

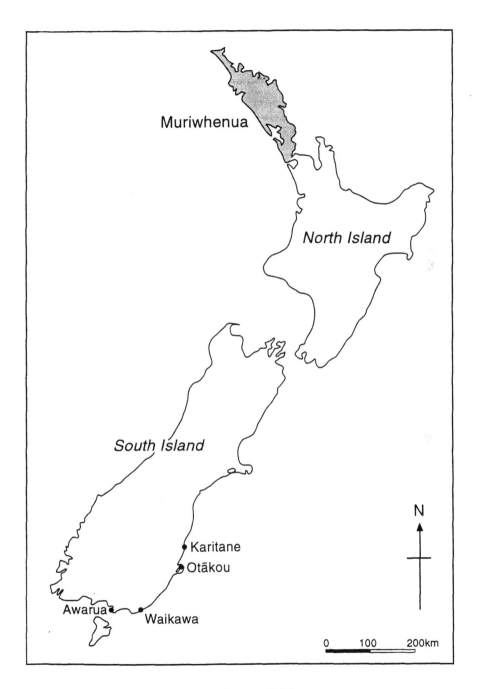

Figure 7.1   Muriwhenua and Kai Tahu fishing areas

incentives, capital grants allowances and tax breaks' (Le Heron and Pawson, 1996: 154; see also Clark *et al.*, 1988: 326; Walker, 1992: 106).

During this period Mäori fishing had re-emerged as a commercial possibility. A number of small-scale Mäori commercial fishers established operations, often combining fishing with another activity like commercial or subsistence farming (Memon and Cullen, 1992: 154). Consequently, as the government removed restrictions on the fishing industry, Mäori were not generally able to participate in the resulting expansion as their activities were at the smaller end of the industry. There were no doubt exceptions, one in particular being Otäkou Fisheries, established after the Second World War, which engaged in both domestic and export trading of fish (Schwimmer, 1957: 32). This operation deserves consideration in more detail as it shaped the national distribution of much Mäori commercial fishing to the present day (see Figure 7.1).

Otäkou Fisheries was a Kai Tahu whänau initiated business (built around a local carrying company), the aim of which was to keep local Mäori working near their families, rather than travelling away to find employment. In 1947 the company exported the first ever whole crayfish to Melbourne and in 1953, on discovering tails were being shipped on to the US, the company initiated direct exports, realising over US$1 million in US sales by 1956 (Schwimmer, 1957: 35). Many whänau who had initially invested in the company were also its suppliers, as independent owner-operators of their vessels (though the company financed a large percentage of the vessels' costs) (Schwimmer, 1957: 36). What ultimately resulted from this initiative was a series of satellite processing plants owned by the company stretching from Awarua (Bluff) to Karitane (a large portion of the South Island's southeast coast). This venture represented a major re-engagement in commercial fishing by one *iwi* - Kai Tahu - after almost a century of having been prevented from such activity. To this day, the greatest concentration of Mäori fishers continues to be between Awarua and Karitane due to the activities of Otäkou Fisheries.

Conversely, North Island Mäori fishers - especially from the East Coast and Muriwhenua - were not so fortunate. According to Memon and Cullen (1996: 255), even for:

> Mäori individuals and tribes who may have had commercial aspirations, the cumulative impacts of decades of marginalisation was a major structural barrier.

This was due to lack of access to funds, as Mäori land was unable to be used for collateral against borrowed capital (Memon and Cullen, 1996: 255).

Mäori land was however appropriated by the government and used for resettlement of returned Päkehä servicemen after both World Wars - often involving the more fertile farmlands under Mäori tenure. The huge casualties among Mäori servicemen in World War Two also created economic pressures for the traditionally labour-intensive Mäori farming systems due to a large reduction in available labour power. At the same time, there was encouragement by the State for Mäori to leave the rural areas and relocate in the cities in order to provide labour in factories, due to the labour shortage resulting from the SecondWorld War.

Consequently, the impact of the SecondWorld War had two contrasting effects on Mäori in relation to commercial fishing. The loss of labour power and the appropriation of land for Päkehä soldier settlements undermined the economics of established Mäori farming systems, thereby creating the pressure for Mäori to diversify into other commercial activities like fishing. However, the erosion of the economic position of Mäori farming and the status of Mäori land also meant that capital to support fishing was scarce. The result was a movement by Mäori into the small-scale end of the fishing industry with many Mäori fishers being highly pluriactive in other commercial and subsistence activities.

The industry that began to develop during the 1960s was strongly influenced by the situation of Mäori. International fishers exploited New Zealand waters for the international market and, contrary to the expectations of the industry's deregulators, had little to do with New Zealand markets. New Zealand-based fishers principally supplied the domestic market, with only a small amount of exporting of crayfish and tinned oysters. This domestic-oriented industry group was also heavily populated by Mäori fishers. While fishing was only a small industry, it provided a haven for Mäori economic activity, particularly as state policy continued to strongly enforce a Päkehä hegemony over pastoral farming and migration to urban factory work for rural Mäori. A futher consequence of the hegemony of pastoral farming in the minds of policymakers was that the fishing industry existed in an arena of significant governmental disinterest over both the degree to which international capital was exploiting the waters around New Zealand and the actual nature and extent of the fisheries resource in New Zealand.

Managed export industry development in New Zealand (1960s-1983)

The New Zealand fishing industry began to experience export growth in the

1970s and early 1980s. The first significant event was that New Zealand joined the international trend and declared a 200-mile exclusive economic zone (EEZ) - one of the largest in the world (Le Heron and Pawson, 1996: 154). Those international fishing companies that had been habitually fishing in the area now included in the zone, were required to engage in joint venture projects with New Zealand companies in order to be able to fish in the EEZ. Consequently, a number of larger companies in New Zealand - Fletcher Challenge, Watties, and Carter Holt Harvey - entered the fishing industry through these joint ventures, while the vast majority of established fishers (predominantly Mäori) remained outside this development. The result was the emergence of a new group of large New Zealand-based companies engaged in fishing and exporting fish through joint venture arrangements. This brought about rapidly increasing problems relating to issues of biological and economic sustainability (Memon and Cullen, 1996: 256), due largely to overcapitalisation of commercial fishing activities (Cullen and Memon, 1990: 54; Kelsey, 1993: 261). The problems were most severe on the North Island's east coast and at Muriwhenua, where 'significant concentrations of Mäori who were part-time commercial' fishers were located (Memon and Cullen, 1996: 255-6). The state, in an effort to stabilise and make the industry more efficient, proposed the adoption of a quota management system (QMS) and individual transferable quota (ITQ) under the 1983 Fisheries Act (Cullen and Memon, 1990: 50; Clark *et al.*, 1988: 327; Dewes, 1989: 131; Kelsey, 1993: 261; Memon and Cullen, 1996: 256). This came into law for offshore fisheries in 1983, and for inshore fisheries (which were far more significant for Mäori) on 1 October 1986.

The QMS was directed at controlling the level of catch taken from a particular fishing region and proposed a number of mechanisms to ensure that overall catch levels fell within the quota for the region. The first aspect was the estimation of a 'total allowable catch' (TAC) to ensure the 'maximium sustainable yield' (MSY) (Le Heron and Pawson, 1996: 155). Within the overall context of commercial and recreational fishing a 'total allowable commercial catch' (TACC) was estimated. These were then administered through regulation of recreational fishers in the form of fishing seasons, size limits and bag limits for some species, and regulation of commercial fishers through the allocation of ITQs (Le Heron and Pawson, 1996: 155).

While all licensed fishers had been issued quotas up until 1986, the new act gave the government the opportunity to undertake a major restructuring of the industry. It was argued by the government that to remove stress from the resource a reduction of fishers would be required. Consequently, up to

1800 fishers were not issued ITQs, and the distribution by MAF greatly favoured the recently arrived joint venture companies at the expense of small fishers (Memon and Cullen, 1992: 160). The fishers who lost ITQs were small, individual, often (though not always) part-time fishers, many of whom were Mäori (Cullen and Memon, 1990: 59). While 85 percent of the 1800 fishers who lost ITQs were Mäori, no Mäori participants in the industry gained any of the remaining ITQs, although Otäkou Fisheries did manage to retain a presence in the industry through a processing deal with Skeggs Fisheries in Dunedin. Instead, the ITQs were granted to a mere eighteen large companies, only some of whom were already participants in the fishing industry (Cullen and Memon, 1990: 59; Memon and Cullen, 1992: 160; Walker, 1992: 107). None of the large companies which made serious applications for ITQs was turned down, as the amount of capital they were able to invest in joint off-shore fishing ventures with foreign companies relieved the state of a potentially enormous monetary burden (Memon and Cullen, 1992: 160). Despite these drastic measures this restructuring reduced the total allowable catch by a mere five percent (Cullen and Memon, 1990: footnotes 28-30, 59).

The intentions of the state in engaging in this restructuring were obscure. The overt reason was to make the fisheries more sustainable, yet a 1983 Ministry of Agriculture and Fisheries study concluded that exclusion of part-time fishers would have a minimal effect on catch reduction. In an obtuse fashion, MAF argued that removing part-time fishers would reduce a potential for expansion from this group, despite MAFs own evidence that profit was related to vessel size for inshore fleets (MAF, 1983: 4). MAF also ignored advice in the Fairgrey Report of 1986 which stated that the social impacts of this action would be devastating, particularly to Mäori of Northland and to other fishers who, after the TAC was lowered, had such small ITQs they were no longer viable (Cullen and Memon, 1990: 59-60; Memon and Cullen, 1992: 160). Such fishers witnessed the large-scale takeover of the industry by large capital as companies bought out their ITQs (Cullen and Memon, 1990: 59-60). A particular barrier for these small fishers was that they were refused bank loans when using ITQs as collateral, while companies could borrow against other assets (Memon and Cullen, 1992: 161).

The result of the ITQ system and the way in which quotas were allocated has a familiar appearance for rural sociologists. At the start of the 1970s, the fishing industry was characterised by a large number of small-scale and often pluriactive fishers engaged in commercial fishing, with another large group also engaged in fishing on a recreational basis. There were few large companies involved and little exporting. By 1987, the ITQ system had entirely eradicated the mass of small fishers in favour of a few large companies. Given that the resulting industry was left with a dual structure of recreational fishers at one end and large corporates at the other, this has a striking resemblence to the 'disappearing middle' observed in the size of farm operations in the US (Buttel, 1983; Buttel and LaRamee, 1991) and New Zealand (Campbell, 1994; Fairweather, 1992). What makes the disappearing middle in the New Zealand fishing industry even more striking is that the vast majority of those who disappeared were Mäori.

The ITQs became an even more contentious aspect of the 1983 Act, particularly after the election of the Fourth Labour Government and the institution of neo-liberal reforms which proposed making the ITQs openly tradeable between companies.

Liberalisation of ITQs and Mäori resistance: From 'Rogernomics' to 'Sealords' (1983-1992)

While the politics surrounding the proposed liberalisation of the 1983 Act in 1986 have been discussed in detail elsewhere (Cullen and Memon, 1990: 50; Clark *et al.*, 1988: 327; Dewes, 1989: 131; Kelsey, 1993: 261; Le Heron and Pawson, 1996: 154-157; Memon and Cullen, 1996: 256), the pivot around which the process unfolded was situated within the issues concerning Mäori that are central to this chapter.

Conflict between Mäori and Päkehä over access to and ownership of marine fisheries, had generally been ignored by government (and might have continued to be so) had not the Third Labour Government (1972-75) created the Waitangi Tribunal to hear Mäori grievances over contemporary claims under the Treaty, and had not the Fourth Labour Government (1984-1990) gone on to amend their earlier legislation to include claims dating back to 1840. Successive governments from 1840 had, as earlier demonstrated, either failed or refused to recognise their obligations to Mäori, in relation to Article II of the Treaty (Kawharu, 1989; Levine and Henare, 1994; Memon and Cullen, 1992, 1996). These claims included the right of Mäori to develop both cultural/ customary and commercial fisheries (Waitangi Tribunal, 1992: 253-256). As this paper has already demonstrated, the relationship between Mäori and fishing had been deep and enduring and significantly predated colonial contact. It was within this context that the liberalisation of the New Zealand fishing industry met with strong resistance. This resistance revolved around both the future status of fisheries under the Treaty and the disastrous consequences for

Māori fishers of the ITQ system.

The legal trigger for a challenge by Muriwhenua to the new fishing industry structure was the 1986 amendment to the act which enabled companies privately to trade their quota. Prior to the ITQ and QMS initiatives, individuals and/or companies competed with each other for fish, yet none owned the resource outright so as to preclude others from it. The ability of companies to trade quota essentially meant that ownership of the resource had passed out of Crown control, and this issue of ownership was central to claims to the Waitangi Tribunal. The Tribunal gave Māori the avenue to lodge a legal challenge to this industry structure at a time when other sectors of New Zealand industry were being swept away by the government's commitment to market forces (known as 'Rogernomics', after the treasurer of the period, Roger Douglas).

The basis of the legal challenge by Muriwhenua was that the issuing of tradeable ITQs ceded tradeable property rights over a fishing resource that Māori had never relinquished under Article II of the Treaty (Walker, 1992:107). The claim was lodged with the Waitangi Tribunal which, betweeen 1986 and 1988 provided government with a series of reports and memoranda (Memon and Cullen, 1996: 257; Walker, 1992: 107). In 1987, Muriwhenua and the New Zealand Māori Council (NZMC) asked the Tribunal for an interim ruling prior to filing an injunction with the High Court against the issuing of ITQs (Walker, 1992: 107). The Tribunal considered a particular area of sea to be a property in the same way as land, and therefore the state must acquire a use right from the traditional owners (Memon and Cullen, 1992: 257; Walker, 1992: 108). The Government ignored this ruling, but High Court Judge Greig concurred with it, ordering an 'interim stop' to the issue of ITQs for Muriwhenua waters (Memon and Cullen, 1992: 162; Walker, 1992: 108). In October 1987, the NZMC, the Tainui Trust Board, the Kai Tahu Māori Trust board and others again went to the High Court, which resulted in a temporary suspension of all ITQs (Memon and Cullen, 1992: 162; Walker, 1992: 108). The advice from both the Tribunal and the High Court was for Māori and government to negotiate via a joint working party of four from each side. The result was a NZ$1.5 million grant without prejudice from the government to the NZMC, who established a Māori Fishing Corporation to negotiate with the Crown over fishing rights (Memon and Cullen, 1992: 163; Walker, 1992: 109). Māori negotiators stated to the Crown that even though under Article II of the Tready they owned 100 percent of the fisheries, they were prepared to share it equally. In 1988 the government introduced the Māori Fisheries Bill which acknowledged Māori fisheries rights, accepted the 50/50 split, but contained little else that met Mäori requests (Walker, 1992: 108). Even this partial gesture created a storm within the industry.

The possibility of a 50/50 fisheries ownership provoked what Walker termed a 'white backlash' (Walker, 1992: 109). This included verbal and media attacks on Mäori and government, culminating in two serious challenges: one from Fletcher Fishing, which threatened to withdraw a NZ$200 million investment in fishing 'because of the uncertainty caused by the Mäori claim'; the other from the National Party, which threatened to repeal the Mäori Fisheries Bill if it came to power in 1990 (Walker, 1992: 110). The government bent under these pressures and withdrew the Bill replacing it with another which gave Mäori 10 percent of the ITQs and a NZ$10 million grant, the remainder of the 90 percent ownership being left in abeyance for the courts to rule upon. It also established a Mäori Fisheries Commission, on which Mäori were effectively outnumbered by government appointed members (Walker, 1992: 110-111). Government further agreed to a moratorium on fishing for remaining species and full hearings of several interrelated Mäori fisheries claims were scheduled for 1990. This action by government partially mollified the concerns of big business, but seriously compromised the Treaty settlement process, leaving it in a difficult situation that was terminated by the loss of the 1990 election.

Good fortune smiled on the state by way of an opportunity that would supposedly appease Mäori fisheries grievances. One of the large fishing companies, Carter Holt Harvey (CHH), was progressively absorbed into foreign ownership, consequently compromising its ownership of Sealords (Kelsey, 1993: 261). CHH had too much quota in foreign ownership, thus breaching the statutory limit, so the state assisted Mäori to purchase a half share of the company, but with conditions attached (Kelsey, 1993: 262-264). One of the most contentious of these conditions from an *iwi* perspective was the revocation of all future claims to fisheries resources under the Treaty. This was agreed to by only some of the Mäori members of the commission, at a time when other Mäori members of the commission were away and unable to have input into this section of the agreement. One interpretation of this hasty decison was that urban Mäori were demanding the deal be settled, believing they would be the group to most benefit from it, because they constituted the largest Mäori grouping (personal communication, 1997). As Davies (1993: 1) states, it was a 'deal full of fish hooks' which continues to be problematic for Mäori.

The aftermath

The Sealords deal, and the allocation of 10 percent of fishing quota to Mäori, had a number of effects. The most serious was the way in which the monies accruing from these new ventures were redistributed among Mäori, effectively establishing the structures around which future Treaty settlements might develop.

There was strong lobbying on behalf of what has been described as a new *iwi* - Ngäti Urban - who represent Mäori who live in cities and have lost touch with their *iwi*. They argued that profits from Mäori fishing quota should be redistributed on a population basis - a formula that would have strongly advantaged urban Mäori. TOKM, however, has been responsible for annually allocating monies from Mäori quota and has respected the former use of fisheries by certain *iwi*. In other words, if a tribe had been marine fishers at contact and had continued to exercise (where possible) or genuinely claim that right up until the issuing of ITQs, then TOKM acknowledges these user rights. The needs of urban Mäori have been addressed by splitting the monies derived from the resource with 60 percent going to *iwi* and 40 percent to urban Mäori. This outcome effectively ensured that *iwi* would be the primary agency for utilising resources regained from future settlements under the Treaty. While fishing has been removed from the claims process, many other resources will continue to fall under the precedents set within the fishing industry.

A more specific concern for fishing *iwi* is that having been removed from participating in the fishing industry since 1986, there is a major problem in re-establishing the type of infrastructure required for new Mäori owners of ITQ. Considerable capital stock was lost to Mäori when they were restructured out of fishing in 1986. Four years later the damage had been done. Another hurdle for *iwi* wishing to re-enter the commercial fishing industry is the way in which the resources have been zoned. For example, ITQs for some species have been allocated to Muriwhenua (Northland) *iwi*, which are located off the West Coast of the South Island, or in the Chathams Bight, southeast of Christchurch. The same scenario could apply to all marine fishing *iwi*, the recent exception being the Moriori *iwi* in the Chathams having been granted sole access rights to the fish in their waters.

Given the loss of capital stock by Mäori fishers, it is not surprising that the *iwi* which has responded most quickly to the new opportunities is Kai Tahu. Otäkou Fisheries retained its presence in the industry and had the capital stock available to re-establish fishing activities more quickly than other Mäori.

In 1997, Kai Tahu established a joint venture with Namibia to fish for orange roughy off the coast of Namibia.

Conclusion

The last 150 years of New Zealand history has witnessed a dramatic series of gains and reversals for Mäori in attempting to participate in one of their most important pre-contact industries - fishing. Mäori fishing skills and the ability to adapt inter-tribal trade to commercial trade with Päkehä settlers was crucial to the economic success of the colony, and during this period Mäori even exported dried fish to Australia. However, with the establishment of a colonial Päkehä economy, Mäori fishers were disenfranchised by the Crown alongside similar disenfranchisement from their land. The Crown's interpretation of the Treaty of Waitangi restricted Mäori to subsistence fishing and eroded the position of Mäori as commercial fishers. From this marginalised position, Mäori emerged again in the years after the Second World War to become the dominant group in the domestic fishing industry, while large international companies were beginning to exploit the off-shore fish resources around New Zealand. As the New Zealand industry began to participate more in the global fish market, and throughout the period when joint ventures were established in the wake of the declaration of the EEZ, Mäori maintained their position of dominance in the domestic market. This was shattered when, from 1983, the state restructured the industry in favour of exporting companies and overseas interests, effectively marginalising Mäori from the New Zealand industry through the ITQ system. In response to this devastating setback, Mäori challenged the legitimacy of the ITQ system through the Waitangi Tribunal, and when the government seemed set to provide redress for the history of disenfranchisement, a severe backlash (aided by corporate capital) heavily undermined any potential gains. Nevertheless, while never being able to reassert the level of resource ownership established under the Treaty, Mäori were able to regain a foothold in the New Zealand industry, and at least one *iwi* has been quick to build on this foothold.

It is tempting to see this as the completion of a full circle, with the progeny of Tuhawaiki - the original exporter of dried fish to Australia - now entering a joint venture with Namibia as global players in the fishing industry. However, such a metaphor suggests a smooth progression or process between starting and end points. Instead, Mäori fishers have been regularly disempowered by the New Zealand state - sometimes in alliance with petty bourgeois settlers and later in alliance with large local and transnational corporates - which has intervened three times to establish and re-establish Päkehä capitalist hegemony over fishing in New Zealand. Each time, Mäori have slowly become re-empowered through the deployment of skills and a tenacious ability to survive against the odds. Given the recent successes of Kai Tahu fishing ventures, the cynical might observe history and suggest that another disempowering intervention by the state is nigh. *Ma te wä ka kite ai tätou* (only time will tell).

Notes

1   A significant exception is the work of Cullen and Memon (1990) and Memon and Cullen (1992; 1996).
2   Henceforth TOKM.
3   The nuclear or extended family; in this case it denotes the latter.
4   Kaik is Kai Tahu dialect for the Northern käinga, or village.
5   Sub-tribes with a founding ancestor from the original tribe, of which it was/is an integral part.

References

Beattie, H. (1994), *Traditional Lifeways of the Southern Mäori: The Otago University Museum Ethnological Project, 1920,* Anderson, A. (ed.), Otago University Press (in association with the Otago Museum): Dunedin.
Broad, T. (1996), *A Discussion Paper,* Otäkou Marae: Dunedin.
Buttel, F. (1983), 'Beyond the Family Farm', in Summers, G. (ed.), *Technology and Social Change in Rural Areas: A Festschrift for Eugene Wilkening,* Westview Press: Boulder, pp. 87-107.
Buttel, F. and LaRamee, P. (1991), 'The 'Disappearing Middle': A Sociological Perspective', in Friedland, W., Busch, L., Buttel, F. and Rudy, A. (eds), *Towards a New Political Economy of Agriculture,* Westview Press: Boulder, pp. 151-169.
Campbell, H. (1994), *Regulation and Crisis in New Zealand Agriculture: The Case of Ashburton County, 1984-1992,* Unpublished Ph.D Thesis, Charles Sturt University: Wagga Wagga.
Clark, I., Major, P. and Mollett, N. (1988), 'Development and Implementation of New Zealand's ITQ Management System', *Marine Resource Economics,* No. 5, pp. 325-349.
Cullen, R. and Memon, A. (1990), 'Impact of the Exclusive Economic Zone in the Management and Utilisation of the New Zealand Fishery Resource', *Pacific Viewpoint,* Vol. 31, No. 1, pp. 44-62.
Davies, J. (1993), 'Sealord: A Deal Full of Fish Hooks', *Overview,* Vol. 46, No. 1, p. 4.
Fairweather, J. (1992), *Agrarian Restructuring in New Zealand,* Research Report No. 213, Agribusiness and Economics Research Unit, Lincoln University: Canterbury.
Graham, D. (1994), *Crown Proposals for the Settlement of Treaty of Waitangi Claims,* Office of Treaty Settlements, Department of Justice: Wellington.
Hargreaves, R. (1960), 'Mäori Agriculture After the Wars (1871-1886)', *Journal of the Polynesian Society,* Vol. 69, pp. 354-367.
Kelsey, J. (1993), *Rolling Back the State,* Bridget Williams Books: Wellington.

Le Heron, R. and Pawson, E. (1996), *Changing Places: New Zealand in the Nineties*, Longman Paul: Auckland.

Levine, H. and Henare, M. (1994), 'Mana Māori Motuhake: Māori Self Determination', *Pacific Viewpoint*, Vol. 35, No. 2, pp. 193-210.

McEvoy, A. (1986), *The Fisherman's Problem: Ecology and Law in the California Fisheries, 1850-1980*, Cambridge University Press: New York.

Memon, A. and Cullen, R. (1992), 'Fisheries Policies and their Impact on the New Zealand Māori', *Marine Resource Economics*, Vol. 7, pp. 153-167.

Memon, A. and Cullen, R. (1996), 'Rehabilitation of Indigenous Fisheries in New Zealand', in Howitt, R., Connell, J. and Hirsch, P. (eds) *Resources, Nations and Indigenous Peoples. Case studies from Australasia, Melanesia and South East Asia*, Oxford University Press: Melbourne, pp. 252-264.

Ministry of Agriculture and Fisheries (1982), *Future Policy for the Inshore Fishery: A Discussion Paper*, Ministry of Agriculture and Fisheries: Wellington.

New Zealand Law Commission (1989), *The Treaty of Waitangi and Māori Fisheries Mataitai: Nga Tikanga Māori me te Tiriti o Waitangi*, Wellington.

New Zealand Māori Council (1976), *Report of the Seminar on Fisheries For Māori Leaders*, Centre for Continuing Education, University of Auckland: Auckland.

Schwimmer, E. (1957), 'Rani Ellison: Māori Crayfish Tycoon', *Te Ao Hou*, No. 20, pp. 32-36.

Taiapa, J. (1996), *Ngā Tikanga-a-Iwi: Māori Society - The Tribal Group*, Massey University: Palmerston North.

Tau, R. (1995), 'A Kai Tahu Perspective on Water', in *Te Karaka*, Ngaai Tahu Publications Ltd: Christchurch.

Te Ohu Kai Moana (1997), *Te Ohu Kai Moana and the Sustainable Use of Renewable Marine Resources - A Discussion Paper on Policy*, Te Ohu Kai Moana: Wellington.

Te Wānanga (1929), Vol. 1, No. 2, Māori Purposes Fund Board: Wellington.

Waitangi Tribunal (1988), *Report of the Waitangi Tribunal on the Muriwhenua Fishing Claim*, Government Printer: Wellington.

Waitangi Tribunal (1992), *The Ngāi Tahu Sea Fisheries Report*, Brooker and Friend: Wellington.

Walker, R. (1992), 'The Treaty of Waitangi and the Fishing Industry', in Deeks, J. and Perry, N. (eds), *Controlling Interests. Business, the State and Society in New Zealand*, Auckland University Press: Auckland, pp. 98-112.

Wheen, N. (1997), *The Treaty Itself - European Constitutional and Legal views of the Rights of Indigenous People*, MAOR420 lecture, March 20.

8 Economic restructuring and neo-liberalism in Australian rural adjustment policy

VAUGHAN HIGGINS

Introduction

An important topic of debate in current agri-food restructuring literature is the extent to which globalisation processes have reduced the autonomy of state action. While it is generally acknowledged that globalisation has undermined the impacts of Keynesian policies (see Buttel, 1996: 26), there is still little research on the role of the nation-state in the agricultural restructuring process. This chapter contributes to knowledge in this area by examining the Rural Adjustment Scheme (RAS), a joint federal-state programme which assists viable producers to become more productive, while providing re-establishment grants for those farming enterprises no longer considered to be viable. The RAS has always been seen by policy-makers in Australia as an instrument for 'managing' change in agriculture. However, what makes the RAS interesting is that since its beginnings in 1977, its philosophy has progressively shifted towards greater farmer self-reliance and individual risk management. This change in focus is most evident from 1988 when the RAS was extensively restructured. The ideological shift between 1977 and 1988 raises some interesting questions concerning the Australian state's involvement in 'managing' structural change. This paper examines the Australian state's involvement in rural adjustment policy up to 1988, and what the move towards a more economically-driven (or 'neo-liberal') policy-making agenda might mean sociologically for capital accumulation in agriculture.

Restructuring and the crisis of Fordism

Changes in rural adjustment policy have not occurred in a social vacuum.

They are related to wider global processes of restructuring and change. To offer an interpretation of the increasing focus by the Australian state on farmer self-reliance, it is useful to use insights from French 'Regulation' theory. The French Regulation School, of whom the best known proponents are Aglietta (1987) and Lipietz (1987; 1992), put forward the idea that in capitalism there is a strong relationship between the economic structure (the *regime of accumulation*) and the socio-political structure (the *mode of regulation*) (Roobeek, 1990). Each mode of regulation is associated with a regime of accumulation which ensures that economic activity continues with as little disruption as possible. It is suggested by Aglietta (1987) and Lipietz (1992) that the postwar mode of accumulation can be labelled as 'Fordism'. The 'intensive' Fordist mode of accumulation and development - carried out under the hegemonic 'leadership' of the United States - focussed on a Taylorist labour process in which mass production was 'matched' with mass consumption. This model of development was disseminated to other advanced capitalist economies such as Australia (see Goodman and Redclift, 1989). According to Lipietz (1992: 6), increased levels of productivity, brought about largely through the development of new technologies after World War Two, necessitated the development of a mass consumption norm so that accumulation would remain stable.

In order to sustain a balance between production and consumption, however, increasing levels of intervention in the labour process were required. This was achieved with modes of regulation - both private and state-based - established after World War Two. While there were some differences between advanced capitalist states, according to Lipietz (1992: 6-7), regulation generally took the form of:

- social legislation covering minimum wage levels and annual wage rises in line with increased productivity;
- a 'welfare state' to ensure that all wage-earners remained consumers, even when they were prevented from earning a living;
- credit money as the economy demanded.

While often neglected in the regulationist literature, individual nation states were important in terms of regulation (see Jessop, 1990). According to Jessop (1994), the Keynesian welfare system, which was established by many post-War Western economies, played a critical role in regulating the Fordist 'accumulation regime'. The focus on linking productivity with purchasing power marked a distinct shift from the previous 'extensive' regime of accumulation, where production occurred with little reference to consumer markets.

The Fordist model of post-war development, however, is considered to have entered a crisis from the mid-1960s. As Marsden (1992: 211) notes, this was provoked by

> the decline in America's imperialist hegemony, the collapse of Bretton Woods in the early 1970s, the oil price rises of the same period, the social barriers placed by organised labour on firms and states to effect substitutions of labour for capital and, probably most importantly, the application of the classical Marxist principle of the structural tendencies within capitalist development for the rate of profit to fall due to the uneven adoption of new technologies.

In Australia, as in other advanced capitalist nations, this situation caused incomes to decrease, rates of productivity to decline and overproduction to become endemic (Fagan and Webber, 1994; Lawrence, 1987). The response by advanced capitalist states to this 'Fordist crisis' was to increase levels of social welfare in order to prevent the collapse of demand. However, a halt in wage rises meant that this could not be sustained. As Lipietz (1992: 16) notes, ultimately the very legitimacy of the welfare-state was called into question, and with it the Fordist compromise. In Australia, the loss of legitimacy was manifested in strong pressures for the abandonment of Keynesian welfare policies from the mid-1970s (Fagan and Webber, 1994).

It is suggested by some (cf. Buttel, 1992; Kenney *et al.*, 1991; Mathews, 1989) that capitalism might now be entering a post-Fordist or neo-liberal era of development with a fundamentally different way of organising capital accumulation. There is much debate concerning the form and functions of post-Fordism, with some arguing that it is questionable whether a new regime of accumulation has even emerged to replace Fordism (see Boyer, 1990; Hampson, 1991; Jessop, 1992). However, there is a general acceptance that decreased regulation by nation-states, structural unemployment, and the growth of neo-liberalism as an ideology, has had considerable social and economic impacts (Buttel, 1992; Marsden, 1992). Increasing global economic integration, in particular, appears to have brought about significant changes in both accumulation and regulation, with the sovereignty of nation states being increasingly eroded (McMichael, 1996: 197).

The growth of neo-liberalism warrants some further attention as it is this view that appears to form the rationale for much state economic restructuring. Neo-liberalism is basically an ideology which stressess that economic resources are best distributed through market forces rather than by government intervention (Stilwell, 1993). It is an ideology which places economic objectives above all else, including social justice. According to Peters (1993), while claiming to provide opportunity for all, this approach effectively marginalises the interests of certain social groups along the lines of gender, ethnicity, class and also, it can be argued, in terms of the rural/urban dimesion. In other words, neo-liberal ideas benefit mainly those who are already privileged with respect to economic wealth and political power.

From a theoretical perspective these points are interesting. If a post-Fordist regime of accumulation has emerged, what role does the state play in its regulation? Equally, what are the trends which point towards a shifting role for the state in regulating capital accumulation? Applied to Australian agriculture and, more specifically, rural adjustment, such questions need to be more specific. For instance, in what ways has the RAS, as a national policy, contributed to agricultural restructuring? What does the shift towards greater economic efficiency measures in the RAS mean for the state's role in agriculture? And, do changes in the RAS signify an altered role for the state in the accumulation process? Using the Rural Adjustment Scheme as a case study, this chapter will begin to examine these questions.

Rural adjustment and the crisis of Fordism

In order to understand the current political rationale for the RAS it is necessary to outline the historical events that shaped its formulation. Such events also provide an important economic and political context for understanding later changes in the scheme. Rural adjustment has been an ongoing feature of Australian agriculture since European settlement first began. State assistance for rural adjustment, however, has only been provided since 1935 (Burdon, 1996: 13), as a response to farm debt brought about through the Great Depression. Contemporary rural adjustment assistance - dating from the early 1970s - needs to be seen as a response to three main crises. These events reflect the beginnings of the Fordist crisis in Australia.

First, the increasing economic importance of manufacturing and mineral development from the late 1960s caused agriculture's share of GDP to decline (Gruen, 1990; Lawrence, 1987). This trend resulted in agricultural economists arguing the need for sector resources to be used more efficiently (Kingma and Samuel, 1977). Second, output-based subsidies, introduced in the 1950s, had caused production to increase rapidly relative to underlying demand conditions, suggesting greater government support would be needed to equalise returns. The rising costs associated with such support caused the rationale for such assistance to be placed under increased scrutiny (Industry Commission, 1996: 18). Finally, as farmers increased their output, the prices they received for their commodities tended to fall relative to the cost of inputs, putting them in a 'cost-price squeeze' (see Lawrence, 1987). While some farmers could afford to invest in capital to become more efficient, and thereby offset their declining terms of trade, many small commodity producers found themselves fighting an uphill battle to survive. Not surprisingly, it was from the late 1960s that a 'low-income farm problem' was beginning to be identified in Australian agriculture (Mauldon and Schapper, 1974; McKay, 1967).

These events signalled a shift in agricultural policy by the state. As the Industry Commission (1996: 18) notes, by the late 1960s governments felt they could no longer maintain assistance based on farmers' costs of production. Policies designed to keep 'unviable' and 'low-income' farmers on the land were seen as counter-productive inhibiting the emergence of a more internationally competitive farm sector. A number of farm reconstruction schemes were set up under the conservative Liberal/Country Party coalition government in the early 1970s as a response to crises in particular industries. These included the Marginal Dairy Farms Reconstruction Scheme (1970), the Fruitgrowing Reconstruction Scheme (1972), and the Rural Reconstruction Scheme (1971). While the first two schemes were aimed at specific industries, the Rural Reconstruction Scheme was more broadly based and can be seen as a response to the downturn in the wheat and wool industries between 1969 and 1971 (Burdon, 1996: 13). Essentially, the Rural Reconstruction Scheme was based on debt reconstruction and farm build-up measures. Rather than supporting farmers to produce more, the state now provided funds to assist producers in amalgamating land and restructuring their debt. While the scheme represented a significant change from the output-based assistance of previous policies, it was not until the Whitlam era that the philosophy underlying rural policy showed a marked shift.

The election of the Whitlam Labor Government in 1972 saw a major change in the way that policy was made and evaluated. Increasingly, policy-making began to be dominated by formal economic approaches, with assistance assessed in terms of its effects on economic efficiency and welfare (Martin, 1990: 156). This 'rationalisation' of the policy-making process is reflected primarily in the establishment of the Industries Assistance Commission (IAC) in 1974. The Whitlam government's desire to rationalise policy-making and assistance measures is also reflected in the commissioning of a major report on agricultural policy (Harris *et al.*, 1974). These changes did not occur by accident, but can be seen as a response to increasing global integration. As Fagan and Webber (1994: 127) note, from this period there was a widespread push for 'deregulation', and the abandonment of a Keynesian policy regime, in advanced capitalist economies.

The so-called Green Paper on rural policy, produced under the Whitlam Government in 1974, had a particular focus on the ways in which governments could effectively intervene in order to improve the workings of the market (Harris *et al.*, 1974). In terms of rural adjustment, the Green Paper states that government assistance for agriculture should 'aim to assist rather than impede market forces' (Harris *et al.*, 1974: 196). Nevertheless, there is also a more 'human' face to the report in that it recognises that there may be justification on welfare and efficiency grounds for governments to 'moderate' the influence of market processes through adjustment measures. The Green Paper is important politically as it questions the idea of 'protection all round'. This is quite different from the dominant way of thinking between the 1940s and 1960s, where market failure was believed to always warrant government intervention (Edwards and Watson, 1978: 200). One can observe a major ideological shift from the Whitlam period towards formal economic analysis in rural policy-making. Such an approach was to become increasingly dominant in later years.

The election of the conservative Liberal-National Country Party Government in late 1975, under the leadership of Malcolm Fraser, saw an increased emphasis on promoting economic efficiency and 'the market'. However, in terms of agriculture, the Fraser government tempered this approach by reviving assistance measures abolished during the Labor period (Watson, 1979: 168), no doubt to ensure that the lifestyle elements of farming were not seriously threatened. Such measures were necessary as the National Country Party had strong ties with the farming sector, with much of its support coming from farmers. Thus, if the Fraser government had challenged ideologies concerning the social importance of family farming, it may have faced a voter backlash.

The issue of rural adjustment was seen as an important priority for the Fraser government, with the IAC asked to report on what measures should be put in place after the Rural Reconstruction Scheme expired in mid-1976 (Burdon, 1996: 14). The IAC found that the previous reconstruction schemes had not provided the necessary opportunities for farmers to adjust to demand-based prices. It recommended that a single, integrated Rural Adjustment

Scheme be put in place under a federal-state agreement, to ensure consistent treatment of farm adjustment across different regions and activities (Industries Assistance Commission, 1976). The government accepted this key recommendation and established the RAS on 1 January 1977, operating under the State Grants (Rural Adjustment) Act (Commonwealth of Australia, 1976).

The RAS and the increasing influence of neo-liberalism

RAS 1977

The RAS was formally created in 1977 under the Fraser Government and brought together a number of previous rural reconstruction measures, many of which have already been discussed. According to the Industries Assistance Commission (IAC) - which strongly recommended introduction of integrated rural adjustment measures - previous reconstruction schemes had not allowed farmers to 'adjust' to demand-based market conditions. 'By not providing opportunities for farmers to adjust to market-determined price incentives, these earlier schemes had not solved the rural adjustment problems' (Industry Commission, 1996: 19). The RAS was thus introduced in order to assist farmers in 'adjusting' to market-determined prices, rather than relying on previous supply-based output support. It was still seen as important, however, to support farmers financially in this 'adjustment' process. According to the then Minister for Primary Industries, Ian Sinclair, the Scheme recognises that government has a responsibility to intervene in the adjustment process:

> ...by providing the financial means to ensure that resources continue to be used in those industries, where their earning power is greatest and, at the same time, provide welfare assistance to those farmers so seriously affected by circumstances that they are unable to remain in the industry (House of Representatives, 1976: 3138).

While the RAS can be seen basically as an economic lever used to dislodge low-income producers from agriculture (see Lawrence, 1987), it is also important to note the continued emphasis on welfare assistance. There was an underlying assumption, particularly by National Country Party members, that it was the responsibility of government at that time to support farmers in times of low income. In fact, the Liberal Member for McMillan, Barry Simon, noted that it was socially desirable that farmers were assured of an income 'to sustain a lifestyle in rural Australia' (House of Representatives, 1976: 3475). This view demonstrated that while a neo-liberal market ideology was beginning to have more prominence in rural policy-making, the social value of farming was still regarded as a crucial consideration. Such a perspective was to have less emphasis in later versions of the RAS.

Under the 1977 RAS Act farmers were to be assisted in 'adjustment' in three main ways. Part A provided assistance to farmers for debt reconstruction, farm build-up and farm improvement; Part B provided carry-on finance for expenses when a rural industry was assessed as having suffered a severe market downturn (but excluding natural disasters); and Part C provided rehabilitation assistance and household support for up to one year to farmers assessed as non-viable, and who were considering 'adjusting' out of the industry. Like later schemes, the 1977 RAS was a joint federal-state operation; the federal government set the policy guidelines and provided the funding, while the states were responsible for the scheme's everyday administration. A new RAS was put in place in 1985, following a review by the IAC, involving the introduction of interest subsidies for parts A and B of the scheme in the place of concessional loans, and the inclusion of assistance to aquaculture. Despite these changes though, the provisions of the 1977 Act remained essentially the same until 1988.

RAS 1988

The RAS was extensively restructured in 1988 by the Hawke Labor Government, following a private consultant's review (see Coopers and Lybrand [W.D. Scott], 1988). A re-orientation of RAS towards improved farm management and productivity, rather than farm family assistance, was a key aspect of this restructuring, with the scheme being administered under new legislation, the States and Northern Territory Grants (Rural Adjustment) Act 1988 (Commonwealth of Australia, 1988). Changes to the RAS also followed major reforms to industry assistance announced in the economic statement delivered by the then-Treasurer, Paul Keating, in May 1988 (Keating, 1988). This statement involved widespread co-ordinated reform across the economy, and can be seen as a response to Australia's continuing balance of payments problems (Martin, 1990: 168). The 1987 stock market crash should also be considered another factor in initiating reform. Capital accumulation was sluggish following the 'crash' and industrial reform could be seen as the means to stimulate profit-making. A further interesting factor in the 1988 reforms was the emphasis on agricultural reform for the purpose of supplying cheap food to Asia-Pacific nations. This involved encouraging the development of a corporate-based, 'high-tech, deregulated, rationalised, vertically-integrated farming model' (Lawrence, 1990: 106). It can be argued that the restructuring of RAS at the same time as agribusiness was being encouraged is no accident. From a sociological perspective one can observe that a restructured RAS was imperative if the state was to send the correct 'signals' to the market, and redistribute resources to those farmers (or agribusiness firms) who would use them in the most 'efficient' manner.

The 1988 RAS represents a significant move towards a more intensely neo-liberal policy-making agenda. According to the Act, the purpose of the scheme is to 'assist in maintaining and improving the efficiency of Australian rural industry and so better place the industry to meet international competition and contribute to the national economy'. This objective can be illustrated by noting some of the comments made by the then Minister for Primary Industries and Energy, John Kerin. Perhaps the defining statement made by the Minister was that, 'there is to be a much stronger emphasis on the notion that farming is first and foremost a business - which must be run along sound business lines if it is to survive and prosper' (House of Representatives, 1988: 1462). This statement marks a decisive shift from previous rural adjustment policies in that it discourages farming to be seen as a family enterprise. Farming as a lifestyle is regarded as secondary to business objectives, with successful farming now being seen as the result of effective risk management. Note the contrast between the quote above from John Kerin, and the comment from Barry Simon, the Member for McMillan, in 1977:

> The retention of farming as a desirable lifestyle will be a result of effective management of farm businesses and effective adjustment, not an objective of it (House of Representatives, 1977: 1465).

An interesting point is that Kerin was actually a farmer, with a degree in agricultural science. In view of his farming background, it is of some significance, to say the least, that he should play down so strongly the lifestyle aspects of farming.

An increasing focus on farming as a business, however, was seen as inevitable, given the factors which were viewed as beyond the control of state regulation:

> Rigid policy responses and smothering administrative arrangements cannot cope with...[the] variability in world commodity prices, exchange rates, interest

rates and climatic conditions (House of Representatives, 1988: 1462).

This comment would seem to indicate that the state saw itself as having progressively less control over the fortunes of Australian agriculture. Also included in the new scheme is more attention to longer term 'management' of adjustment problems rather than short-term 'quick-fix' solutions, with more managerial and financial responsibility being devolved to the states. Presumably the rationale for this was that the states would be able to tailor RAS assistance more specifically to individual farm enterprises.

The 1988 RAS was based on the following format:

- A Farm Financial Management Skills Program was added to Part A to provide for improvements in farmers' managerial and technical skills;
- Drought assistance was added to the carry-on assistance provision in Part B of the scheme, coinciding with the removal of drought from the register of national disasters. A revised assessment criterion allowed assistance to be provided in circumstances where farmers face financial difficulties due to circumstances beyond their control;
- The rehabilitation grant was replaced with a re-establishment grant (A$28,000 grant indexed to the consumer price index) under Part C to be made available to eligible farmers leaving the industry. The duration of farm household support assistance was reduced from three years to two.

An important aspect of the 1988 RAS is its emphasis on education and training as a means to 'improve the risk management skills of farmers'. By improving farmer's knowledge and skills it is assumed that they will manage their risks more effectively, and thereby be more productive. This, however, is a very individualistic view of farming and ignores the fact that most of the problems Australian farmers face are structural in nature (Lawrence, 1987). Nevertheless, a political strategy of self-reliance appears a legitimate means for the state to distance itself from the problems facing farm families. Thus, if a producer fails in farming the individual is deemed responsible for the failure - not the state (Lawrence, 1987: 201). Since 1988 the RAS has been reviewed twice with an increasing emphasis on productivity and individual risk management. Most importantly, in 1992, Farm Household Support was removed from the scheme to be administered under the Department of Social Security. While the changes since 1988 are significant, there is not the space to discuss them within this chapter.

Conclusion

What do the changes in the RAS say about the Australian state's role in agriculture, and capital accumulation in general? It is clear that the state's role in agriculture has shifted since the early 1970s, from widespread farm assistance towards selective intervention measures. An increasing focus in public policy on 'self-reliance' and 'individual risk management' confirms the view that while the state is continuing to intervene in the agricultural accumulation process, such intervention is aimed ultimately towards restructuring, and making the market work more 'efficiently' in the interests of larger-scale farmers and agribusiness. The increasing emphasis on farming as a business, rather than as a lifestyle, also represents part of this trend. Do these changes in policy represent a fundamentally different era in terms of the Australian state's role in organising capital accumulation, or are they part of a struggle by the state to 'manage' a continuing Fordist crisis? On analysing the evidence, it appears that the state's role in accumulation certainly shifted from 1988. While the state has become more neo-liberal in its approach to policy-making since the 1970s, it was not until 1988 that there was a major shift away from a Keynesian-based welfare philosophy. The state played a fundamental role in capital accumulation in the 1970s, although this was tempered with an acknowledgment that social and welfare aspects of farming were integral to effective policy. RAS 1988 represents a radical departure from this idea in that farming is seen almost totally as a business entity, with any long-term welfare assistance seen as sending the 'wrong' signals to producers. However, this change by no means represents a shift beyond the Fordist crisis, as it remains unclear how effective a neo-liberal strategy will be in terms of regulation. That is to say, further research is required to ascertain whether a neo-liberalist policy regime - which favours restructuring in the interests of larger scale capital - is capable of providing a long-term stable environment for capital accumulation in agriculture.

References

Aglietta, M. (1987), *A Theory of Capitalist Regulation: The US Experience*, Verso: London.

Boyer, R. (1990), *The Regulation School: A Critical Introduction*, (C. Charney, Trans), Columbia University Press: New York.

Burdon, A. (1996), 'Commonwealth Government Assistance for Adjustment in Agriculture' in Department of the Parliamentary Library (eds), *Australian Rural Policy Papers 1990-95*, Australian Government Publishing Service: Canberra, pp. 1-65.

Buttel, F. (1992), 'Environmentalism: Origins, Processes and Implications for Rural Social Change', *Rural Sociology*, Vol. 57, No. 1, pp. 1-27.

Buttel, F. (1996), 'Theoretical Issues in Global Agri-food Restructuring', in Burch, D., Rickson, R. and Lawrence, G. (eds), *Globalisation and Agri-food Restructuring: Perspectives from the Australasia Region*, Avebury: Aldershot, pp. 17-44.

Commonwealth of Australia (1976), *States Grants (Rural Adjustment) Act*, Australian Government Publishing Service: Canberra.

Commonwealth of Australia (1988), *States and Northern Territory Grants (Rural Adjustment) Act*, Australian Government Publishing Service: Canberra.

Coopers and Lybrand [W.D. Scott] (1988), *Review of the Rural Adjustment Scheme*, Department of Primary Industries and Energy: Canberra.

Edwards, G. and Watson, A. (1978), 'Agricultural Policy', in Gruen, F. (ed.), *Surveys of Australian Economics*, Allen and Unwin: Sydney, pp. 187-239.

Fagan, R. and Webber, M. (1994), *Global Restructuring: The Australian Experience*, Oxford University Press: Melbourne.

Goodman, D. and Redclift, M. (1989), 'Introduction: The International Farm Crisis', in Goodman, D. and Redclift, M. (eds), *The International Farm Crisis*, Macmillan: London.

Gruen, F. (1990), 'Economic Development and Agriculture since 1945', in Williams, D. (ed.), *Agriculture in the Australian Economy*, Sydney University Press: Sydney.

Hampson, I. (1991), 'Post-Fordism, the 'French Regulation School', and the Work of John Mathews', *Journal of Australian Political Economy*, No. 28, pp. 93-128.

Harris, S., Crawford, J., Gruen, F. and Honan, N. (1974), *Rural Policy in Australia*, Australian Government Publishing Service: Canberra.

House of Representatives (1977), *Hansard: Debates (Commonwealth of Australia)*, Vol. 102.

House of Representatives (1988), *Hansard: Debates (Commonwealth of Australia)*, Vol. 163.

Industries Assistance Commission (1976), *Rural Reconstruction*, Report No.76, Australian Government Publishing Service: Canberra.

Industry Commission (1996), *Industry Commission Submission to the Mid-Term Review of the Rural Adjustment Scheme*, Australian Government Publishing Service: Canberra.

Jessop, B. (1990), 'Regulation Theories in Retrospect and Prospect', *Economy and Society*, Vol. 19, No. 2, pp. 153-216.

Jessop, B. (1992), 'Fordism and Post-Fordism: A Critical Reformulation', in Storper, M. and Scott, A. (eds), *Pathways to Industrialisation and Regional Development*, Routledge: London, pp. 46-69.

Jessop, B. (1994), 'Post-Fordism and the State', in Amin, A. (ed.), *Post-Fordism: A Reader*, Blackwell: Oxford, pp. 251-79.

Keating, P. (1988), *May Economic Statement*, AGPS: Canberra.

Kenney, M., Lobao, L., Curry, J. and Goe, W. (1991), 'Agriculture in U.S. Fordism: The Integration of the Productive Consumer', in Friedland, W., Busch, L., Buttel, F. and Rudy, A. (eds), *Towards a New Political Economy of Agriculture*, Westview: Boulder, pp. 173-188.

Kingma, O. and Samuel, S. (1977), 'An Economic Perspective of Structural Adjustment in the Rural Sector', *Quarterly Review of Agricultural Economics*, Vol. 30, No. 3, pp. 201-215.

Lawrence, G. (1987), *Capitalism and the Countryside: The Rural Crisis in Australia*, Pluto Press: Sydney.

Lipietz, A. (1987), *Mirages and Miracles*, New Left Books: London.

Lipietz, A. (1992), *Towards a New Economic Order: Postfordism, Ecology and Democracy*, Oxford University Press: New York.

Martin, W. (1990), 'Rural Policy', in Jennet, C. and Stewart, R. (eds), *Hawke and Australian Public Policy: Consensus and Restructuring*, Macmillan: Melbourne, pp. 155-179.

Mathews, J. (1989), *Tools of Change*, Pluto Press: Melbourne.

Mauldon, R. and Schapper, H. (1974), *Australian Farmers Under Stress in Prosperity and Recession*, University of Western Australia Press: Perth.

McKay, D. (1967), 'The Small Farm Problem in Australia', *Australian Journal of Agricultural Economics*, Vol. 11, No. 2, pp. 115-132.

McMichael, P. (1996), *Development and Social Change: A Global Perspective*, Pine Forge Press: Thousand Oaks.

Peters, M. (1993), 'Welfare and the Future of Community: The New Zealand Experiment', in Rees, S., Rodley, G. and Stilwell, F. (eds), *Beyond the Market: Alternatives to Economic Rationalism*, Pluto Press: Sydney, pp. 171-185.

Roobeek, A. (1990), *Beyond the Technology Race: An Analysis of Technology Policy in Seven Industrial Countries*, Elsevier: Amsterdam.

Stilwell, F. (1993), 'Economic Rationalism: Sound Foundations for Policy?' in Rees, S., Rodley, G. and Stilwell, F. (eds), *Beyond the Market: Alternatives to Economic Rationalism*, Pluto Press: Sydney, pp. 27-37.

Watson, A. (1979), 'Rural Policies' in Patience, A. and Head, B. (eds), *From Whitlam to Fraser: Reform and Reaction in Australian Politics*, Oxford University Press: Melbourne, pp. 152-172.

9 The restructuring of industry-based agricultural training in New Zealand

RUTH SCHICK AND RUTH LIEPINS

Introduction

The election of a Labour government in 1984 signalled a period of major restructuring for both agriculture and the wider political economy of New Zealand (Dalziel and Lattimore, 1996). This involved a rethinking of the relationships between the economy, civil society and the state. The impact of these changes on agriculture and rural society in New Zealand was considerable, and has formed the subject of a growing body of literature by rural researchers since that time. Prior research into the issue of agricultural labour has tended to concentrate on three aspects of change affecting such labour in New Zealand:

- changing patterns of work in farm households - engaging in pluriactivity, off-farm work, substitution of family for paid labour or self-exploitation by family labour (Benediktsson *et al.*, 1990; Campbell, 1994; Keating, 1994; Keating and Little, 1991; LeHeron *et al.*, 1991; Taylor and Little, 1995; Taylor *et al.*, 1997; Wilson, 1994).
- the changing pattern of womens' participation in agricultural and rural paid labour (Anderson, 1993; Pomeroy, 1988; Rivers, 1992; Taylor and Little, 1995).
- the changing position of hired farm labour *vis-a-vis* family labour contributions (Benediktsson, *et al.*, 1990; Campbell, 1994; Fairweather, 1989; 1995).

None of this work has extensively considered either the changing skills content of rural labour or the way in which state restructuring of educational providers has impacted on agricultural production.

The restructuring of the funding, accreditation and provision of educational services, which included those servicing the agricultural labour market, was a key component of reforms after 1990. After 1984, the drastic decline in the financial performance of farms put increasing pressure on farm families to reduce their expenditure on farm inputs, of which a major component was paid labour (Campbell, 1994). Without wishing to review in depth the overall nature and consequences of the agricultural crisis that followed 1984, this chapter will examine one previously unresearched aspect of the changing relationship between the New Zealand state and the agricultural labour process; namely, the way in which state provision of skills to agricultural labour has been significantly restructured during the 1990s, over the period of the 'second wave' of economic and social restructuring that was implemented by the National government of 1990. National government policies from 1989 extended the educational reforms introduced by the Fourth Labour Government, with the intention of firmly establishing a market-led system of education (Lauder, 1991). This has included policies intended to introduce 'market driven' system of vocational training across all industry sectors (Education and Training Support Agency, 1993a; 1994; Smelt, 1995).

Vocational training and New Zealand policy

Internationally, vocational training has been the focus of a variety of ideological, policy and practical efforts for many years. It is viewed as a form of education which offers targeted inputs to the economy, and is less concerned than general education with the attitudes and orientations towards citizenship and participation in civil society (Addison and Siebert, 1994; Cohn, 1979; Cutler, 1992; Kearns, 1993; Purves, 1988).

In 1992, following the passage of the Industry Training Act, a new set of inter-organisational and funding arrangements were put in place in New Zealand, regulating, in particular, workplace and apprenticeship training, including agricultural training up to entry tertiary level. Skill New Zealand, sometimes referred to as the 'Industry Skills Training Strategy', is the name given to the umbrella 'strategy', which since 1992 has provided the justfication and conceptual coherence for the organisation and funding of 'industry driven' agricultural training (Education and Training Support Agency, 1993a; Education and Training Support Agency and New Zealand National Qualifications Authority, 1994; Smelt, 1995). Its goals therefore, provide the criteria against which any policy-oriented evaluation of the programmes and practices of industry-based training might be assessed. No single document outlines the objectives or rationales of Skill New Zealand, but from 1993 such statements appear in a wide range of publications by the Education Training and Support Agency and the National Qualifications Authority (Education and Training Support Agency, 1993a; 1996f; Education and Training Support Agency and National Qualifications Authority, 1994; New Zealand Qualifications Authority, 1991; 1993; 1996a; 1996b).

References to Skill New Zealand identify the relationships between education and the economy as a central priority with the government, citing education, in particular, as a crucial component in ensuring the competitiveness of the New Zealand economy. However, the political nature of education as a state service and the particular history of educational politics in New Zealand society, is also powerful in shaping the rhetoric of educational reform (Codd, 1990; Kearns, 1993; Lauder, 1991; Olssen and Matthews, 1997). Thus, policy documents draw both on discourses of economic productivity (usually framed in terms of 'human capital' or 'market driven' rhetoric) and discourses of social equality. Consistent with other reforms affecting agriculture, they mark a shift in policy emphasis towards economic perspectives which emphasise 'choice' and 'efficiency', system rationalisation and economic viability. In education, these target both operations of educational services and orientations towards education more broadly (Codd *et al.*, 1990; Education and Training Support Agency, 1996f; Education and Training Support Agency and New Zealand National Qualifications Authority, 1994; Olssen and Matthews, 1997). On the other hand, as a social service and one that is historically, ideologically and legally bound to notions of equity and opportunity, these same policy documents employ narratives of equity, access and opportunity.

Skill New Zealand stressed that education needed to be 'relevant' to industry. The government's publications suggested that freed of government restrictions, the market would provide employers and workers, including trainers and trainees, with the information necessary to support the free movement - and subsequently efficient allocation - of capital and skills. This would then boost efficiency and productivity, both of particular industries and of the economy as a whole. It would also increase the employment opportunities of trainee/workers, including those of members of historically marginalised groups. To support the development of the 'training' culture in the New Zealand economy, the government provided subsidies to training and support for the development of the relevant training organisations (Education and Training Support Agency, 1994; 1995; 1996a; Smelt, 1995).

While the key words used in government statements, such as 'access' and 'inclusiveness', 'quality', 'flexibility' and 'portability of qualification', seemed to reflect a concern with social equality, they also mimicked the national direction towards a liberalisation of the economy, and a removal of government regulation and direction, apparently enhancing the power of individuals to make 'informed' economic decisions (Lauder, 1991; Smelt, 1995). The new initiatives would serve new industries, established industries without a prior history of nationally recognised vocational training, as well as industries which had participated in the prior apprenticeship system (Dickson, 1996). They would increase the participation rates of groups of student who had previously not gained access to training (Education and Training Support Agency, 1993b). They would permit the recognition of trainees' prior learning (from non-formal training and work experience) and facilitate the development of training in new content areas (New Zealand Qualifications Authority, 1991). The new system would support a variety of providers and forms of educational provision and a variety of forms of assessment of learning. Qualifications gained through training were to be portable in that they would be nationally registered and thus recognised not only by individual employers, but also within and across industries nationally. The quality of training would be addressed through an industry-led, nationally moderated system of curricular planning; the national registration of units of training; and national certification of assessors of trainee achievement (Education and Training Support Agency and New Zealand National Qualifications Authority, 1994; New Zealand Employers Federation, 1996; New Zealand Qualifications Authority, 1991; 1996a).

Concern with balancing government regulation and subsidisation on the one hand, with market relevance on the other, is central to the new industry training scheme (Smelt, 1995). Within the training scheme, industry accountability to government is accomplished through relationships between industry training organisations (ITOs) and two key government organisations - the National Qualifications Authority (NZQA) and the Education and Training Support Agency (ETSA). Government regulations determine the eligibility of industry training organisations for government training subsidies (Education and Training Support Agency, 1996b; 1996c; 1996d; Smelt, 1995).

ETSA administers government-contestable funding subsidies to industry training. It also determines the eligibility of organisations to be registered as Industry Training Organisations, entitling then to compete with each other for industry training subsidies. The receipt of funds from ETSA occurs only when ITOs make good on projected participation rates, which are registered by formal training contracts signed by trainees and employers (Durham, 1996; Smith, 1997). ETSA's performance in relation to industry training is evaluated by government, and is based on the number of recognised ITOs, the number of ITOs with registered qualifications, and the numbers of training contracts per ITO (Education and Training Support Agency, 1996e).

It is predominantly by adhering to NZQA guidelines and participating in the NZQA mediated processes of degree development that industry developed training is eligible for government support. The NZQA oversees and moderates the development of national qualifications and their component bits ('unit standards') by ITOs, and in some cases, independent industry advisory groups (New Zealand Employers Federation, 1996). NZQA also registers training providers and assessors, keeps student records on participation in training, on achievement of unit standards and on degree status (New Zealand Employers Federation Inc., 1996). NZQA performance in relation to industry training is evaluated by government on the basis of growth in the numbers of registered qualifications and unit standards, providers, and the number of students participating in framework registered programmes of training (New Zealand Qualifications Authority, 1997).

ETSA and the NZQA both have enormous power under this scheme. Their application and reporting requirements are the clearest indication of the operational, as opposed to rhetorical, emphasis of these reforms. ITO performance is evaluated, first and foremost, on the basis of the number of training contracts signed, and secondarily on the basis of the number of qualifications developed and registered and the participation in the industry of members of traditionally marginal groups (Education and Training Support Agency 1996b; 1996c; 1996d).

On the market side, industry leadership is accomplished through the contractual relationships between ITOs, education providers, employers and trainees (Education and Training Support Agency, 1993b; 1996f; Education and Training Support Agency and New Zealand National Qualifications Authority, 1994).

The Agriculture Industry Training Organisation

Le Heron (1992) argues that since the Second World War, New Zealand had participated in a stable global regime of agrifood relations. Since 1973, this regime has become increasingly fractured, with the agricultural sector experiencing a period of prolonged instability, which accordingly influenced demands for agricultural labour. These fluctuating demands for agricultural labour have continued into the 1990s (Campbell, 1994). This period has witnessed shifts in the size of New Zealand agriculture, in the products and organisation of production, and the number and size of farms (Le Heron, 1996). Both the proportion of agriculture involved in production of food related products and the types of products has changed, particularly in relation to pastoral sheep farming. The profitability of agriculture has declined markedly, in part as a function of shifts in the prices for agricultural commodities, fluctuations in the value of the New Zealand dollar, and in part as a result of the retrenchment of government subsidisation, regulation and protection of agriculture (Campbell, 1994; Cloke, 1989; Cloke and LeHeron, 1994; Dalziel and Lattimore, 1996; Fairweather, 1989; Fairweather, 1992; LeHeron, 1996).

As farmers face ongoing pressure to increase production efficiencies, and express this in their changing demands for agricultural labour, these trends in agriculture involve a significant shift in the types and number of skills required for the agricultural labour force (Lawrence, 1987; Taylor and Little, 1995; Willis, 1991). Declining rates of employment in agriculture (Dalziel and Lattimore, 1996; Willis, 1991) also suggest that employment in agriculture is not as likely a prospect for those entering the labour force, or as secure a prospect for those already in the labour force. Lastly, a notable feature of agriculture in New Zealand is the historical pattern of labour market segregation by race and gender.

It is in this context that the Agriculture Industry Training Organisation (agITO) was granted accreditation by ETSA in 1995. In keeping with the aims of Skill New Zealand, the training brokered by the agITO should provide the means by which those already employed in agriculture can up grade their skills. It should thereby both increase the efficiency and productivity of agriculture, and increase the employability of those in the agricultural labour force.

The issue in regard to training provided through the agITO is whether it can actually meet the objectives of Skill New Zealand. Two sets of issues are relevant here. The first is whether the structures and practices of agITO-brokered training are appropriate to the actual needs of agriculture in New Zealand. The second is how the current forms of agITO training perform in light of the research literature on vocational and agricultural training. This research examines the outcomes of vocational education, relative both to other forms of training and to the economic claims made for vocational training. Could the systems of training brokered through the agITO, under any circumstances, meet the objectives of Skill New Zealand?

Equity and efficiency: Does agITO training achieve the objectives of Skill New Zealand?

The agITO central organisation consists of a five-tiered structure:

- a National Board, Finance Committee, and Industry Advisory group, which liaise with government;
- Industry Funding Boards (consisting of stock and station, game, pork, poultry etc.);
- sector specific Training Committees corresponding to these funding boards;
- regional training committees; liasing with,
- regional field operations (Agriculture Industry Training Organisation, 1996).

Between 1995 and 1996, the agITO grew significantly. Operating revenue increased by over NZ$500,000 (Agriculture Industry Training Organisation, 1996), and trainee numbers increased from 1,869 in June 1996 to 3,252 by June 1997 (Education and Training Support Agency 1997a; 1997b).[1] Between 1996 and 1997 the number of registered qualifications increased, as did industry financial contributions (Durham, 1996; Agriculture Industry Training Organisation, 1996). Patterns of participation on training have continued throughout the history of the ITO to be segregated by ethnicity and gender. In June 1997, participation rates by ethnicity and gender in agITO brokered training were 90 percent Päkehä (78 percent male/ 10 percent female) and 9.5 percent Mäori (seven percent male/two percent female) (Education and Training Support Agency, 1997a; 1997b).

The list of qualifications being developed suggest that the ITO is indeed representing a broader range of occupations than were previously served through government training schemes. Increased rates of industry contribution to ITO operating costs suggest that the qualifications it is developing are in fact relevant to the needs of the agricultural industry (Agriculture Industry Training Organisation, 1996). This suggests that the training being developed is of sufficient value to the farmer/employers that they are willing to meet some of the costs of training their employees. This is consistent with the government's intention of providing the subsidies to training essentially as an incentive to industries to become involved and to expand training, which would later become 'self supporting' - i.e. 'industry funded' (Smelt, 1995).

However, little or no information is available which informs us of the value of training in meeting Skill New Zealand's objective of enhancing productivity (in this case agricultural productivity), or trainees' employment opportunities. Such information, based on interviews with employers or trainees, might involve tracing the patterns of their future employment and training activities. The 1996 agITO Annual Report cites a survey of employers who claim to be very satisfied with training, but provides little information on sampling, sector, type of employer, or the questions employers were asked regarding productivity or wage effects of training (Agriculture Industry Training Organisation, 1996). Needs analyses regarding the amount or type of skills needed in the agricultural sector in general or in relation to areas of agITO training specifically do not seem to be informing decision making.

Participation data show little change in the level of inclusiveness of Māori or women in agricultural training (Education and Training Support Agency, 1997b). The main forms of participation data available are head counts by the provider (New Zealand Qualifications Authority, 1997). Data for individual providers, for units standards and for qualifications are considered 'commercially sensitive' (Durham, 1996). The absence of data on completion rates or success rates at assessment for unit standards, means that there is no basis for comparison across different demographic groups, occupational areas, or training providers within agriculture. Somewhat more detailed participation information is provided in a 1996 commissioned study and suggests that indeed completion rates are extraordinarily low (Manthei *et al.*, 1996). One possibility is that trainees do not perceive training to be enhancing their labour market position. The fact that ITO funding is reliant on subsidies which are allocated on the basis of the number of training contracts signed (rather than on any assessment of completion rates, employment rates, or labour market surveys of employers), suggests that at least in this case, the government subsidies are in fact operating at cross-purposes to the stated objectives of Skill New Zealand.

The absence of relevant and systematic data concerning current and projected demands for skills in the agricultural labour market leads us to question how training (both that which is currently available and that which is being developed) will contribute to productivity or efficiencies in agricultural production, how it could affect worker mobility, or equality of employment in agriculture. While government regulation, organisational support and accountability measures were intended to assist industries to develop a training culture, they did not include support for labour market analyses or for training programme evaluation and follow-up. There was no reason to suppose that industries would possess the perspectives or expertise to conduct such analyses. In the absence of such data it is difficult to envision how the training scheme might deliver on the objectives of Skill New Zealand, or alternatively, how one might determine if it had (Kearns, 1993).

Issues from the literature on education and economic productivity

A major concern regarding the contribution of education to agriculture is how best to inform and support farmers' decision making to encourage the most efficient organisation of agricultural production (in relation to any chosen outcome) (Moock and Addous, 1995; Winter, 1997). Skills provided through training will only affect agricultural production if these are the skills which farmers are willing to purchase. This is relevant to concerns with vocational training for two reasons: the first involves the ways in which education or training influence the decision making of farmers/employers regarding investment in labour as opposed to other farm inputs; the second involves the manner in which the provision of training affects decisions by workers to participate, and consequently the skill levels and mix, in the labour force and the productivity of labour (Min, 1995).

The design and evaluation of a vocational training system, the stated objectives of which are to enhance productivity and national competitiveness, would in theory need to attend both the micro-level assessment of the effectiveness and efficiency of alternative forms of training recruitment, provision and delivery, as well as the macro-level patterns of demand for the skills produced (Caillods, 1994; Higgins, 1994; Kearns, 1993; Min, 1995; National Advisory Council on the Employment of Women, 1993). The latter evaluation would require a complex modelling of the economic and social relationships necessary to production as well as considerations of the types and rate of change and risk factors involved, and the appropriate collection and analysis of data on an ongoing basis (Addison and Siebert, 1994; Caillods, 1994; Kearns, 1993).

Reviews of research on the contribution of education to economic productivity point to conceptual problems in the definitions and measurements of education and of productivity, and in the modelling of their relationships. Such problems are also found in the design and operations of the Skill New Zealand scheme (Carnoy, 1995). The research which has looked at the contribution of education to agricultural productivity specifically, has focussed primarily on measures of productivity associated with farms' gross revenue rather than with the practices and decision-making associated with running and working on farms (Moock and Addous, 1995). This offers little to inform us about the processes by which education or human capital more generally, contribute to agriculture. Farmers' assessments of the relative efficiency of various inputs (e.g. technology or labour) and the availability of unpaid family labour which can be drawn on in times of economic austerity will affect the degree to which any training programme impacts on patterns of agricultural employment; the market value of agricultural credentials; and the productivity effects of increased provision of agricultural training. Research which has looked to the farm manager has not always recognised the complexities of the farm decision-making process and the way this is structured by gender, age, property ownership and type of labour contribution (Liepins and Campbell, 1997; Winter, 1997).

Training programs organised around false assumptions about the social organisation of employer and worker decision-making will have a more limited impact than programmes which are appropriately informed (Higgins, 1994; National Advisory Council on the Employment of Women, 1993). Gender - as it relates to the organisation of decision-making on farms - is of particular significance in relation to the participation of females in agricultural training, their recognition as farmers and their participation on the training boards which are designing the content and delivery of agricultural industry training. Further, a focus on individuals as decision makers which does not account for the social organisation of decision making, again obscures the situation in which the educational levels and communication in the farm family, rather than the educational level of a nominal individual head of the family farm, shapes decision making regarding farm management (Min, 1995; Moock and Addous, 1995; Winter, 1997).

A key feature of much of the research on the contribution of education to economic productivity, also central to the New Zealand approach to vocational training, is an emphasis on an 'individual productivity function' (Min, 1995). The 'individual productivity function' supposes that individual productivity is shaped by the fit between worker characteristics, job skill requirements and the reward and promotion incentives in particular jobs. Training - in particular training based on industry knowledge of skill needs - can address a mismatch among these by altering workers' skills to improve their match with the skills required in available jobs. In theory, this will increase both the employability and the productivity of workers.

The research which addresses the relative contribution of vocational-versus-general education has compared the relationships between education (measured in terms of proxy measures for education in terms of years of education completed) and productivity (measured in terms of wages earned)[2] for each type of education (Min, 1995). Most studies have taken a cross-sectional snapshot approach rather than a longitudinal one (Min, 1995) and have rarely addressed the relative rates of employment (including duration of unemployment) for those with vocational as compared to general education; and the relative rates of employment for those who have received vocational as compared to general education *in the area for which vocational training is received*. Overall, these studies have shown that relationships between education and productivity depend on a host of factors including programme characteristics, student characteristics (including gender and race) and the types of training and training institutions (Min, 1995).

An additional consideration in examining the contribution of education to agricultural productivity is the way in which added education affects the wages and employment rates of members for social groups which have historically been marginalised with relation to particular occupations and industries (Addison and Siebert, 1994; Caillods, 1994; Higgins, 1994; Knoke and Ishio, 1994). These questions tend to be treated in a body of literature which attends specifically to segmented and internal labour markets (Carnoy, 1996a; 1996b; Carnoy and Gong, 1996; DeFreitas, 1995; Doeringer, 1995; Knoke and Ishio, 1994). Segmentation of labour by gender and race within agricultural labour markets will also impact on the degree to which changing skill patterns among women and non-Päkehä will increase their participation in the agricultural labour force and therefore their employability or wage rises, or agricultural production processes. Recognition of these issues suggests that increasing access to training will not in itself guarantee the improved labour market position or productivity of members of historically marginal groups relative to members of majority groups. Rather, a variety of factors including attitudes of co-workers, the culture of the workplace and individual as well as sectoral hiring patterns, will impact the relative market value of vocational education for members of different groups.

This research has produced no consistent evidence showing vocational training to be superior to general education in improving productivity. It tends to beg the question of what constitutes 'productivity' and hence of the relationship between education and productivity. It shows that the mix of factors which inform the decision making of employers and workers, rather than simple proxy measures, such as only measuring years of education and wages, for the contribution of education to productivity, are essential to assessing the value of a training programme in relation to any particular set of goals. An emphasis on the skill levels of individual workers has minimised the consideration given to social 'skills' and to group level phenomena (such as the social organisation of the workplace) on farmers' and workers' decision making, on equality in the work force, and on agricultural productivity.

The research suggests quite strongly that the involvement of industry in the design and provision of worker training is in itself insufficient to guarantee the relevance of training, its contribution to equality, or to industry productivity. Whether training is industry or government led, data adequate to assessing the factors which affect both worker and employer decision making are crucial to evaluating the demand for skills and their contribution to productivity and equality (Caillods, 1994; Kearns, 1993; Min, 1995). Government subsidies and organisational support to industry to develop training schemes under the rubric of Skill New Zealand, could conceivably support such data collection and the development of informed decision making in this area. But the tailoring of such efforts to the conditions of different industries would be essential. Efforts to ensure that the data collected are relevant and meaningful in assessing the demand for skills and the value of training qualifications in the labour market, would seem to be far more important both to industry and to evaluating the success of policy, than the current practice of recording the expansion of rates of entry into training, without evaluation of subsequent completion rates or employment experiences of trainees (Caillods, 1994; Kearns, 1993).

Conclusion: The implications for the agri-food system

In New Zealand, as elsewhere, educational reforms impacting on agricultural training have been justified in relation to generic rationales regarding the value and organisation of education in relation to the economy (Addison and Siebert, 1994; Caillods, 1994; Cutler, 1992; Winter, 1997). As such, the reforms affecting agriculture often have not explicitly been differentiated from those affecting other sectors. Shifts in attitudes towards the role of education generally, and the government's role in post-compulsory education specifically, have had particular effects on the access, design, funding and economic value of agriculture-related training. In spite of government rhetoric and the initial steps in the direction of industry-led and industry-supported training, the pressures faced by agriculture in New Zealand have included a contraction of labour inputs in farming which educational policies can do little to mitigate. The features of training funding and provision, such as subsidies for trainee wages and subsidies to farmers for providing site based training, can have some effect on the affordability both of labour and farmers' participation in farm-based training, but expansion of participation in training does not in itself produce results in the areas of industry productivity or competitiveness, employment rates, or equality of employment (Addison and Siebert, 1994).

The ideological drive behind the reforms to vocational training in New Zealand placed employers and workers in the driver's seat to allow economic, industry and labour market considerations rather than government agendas to drive the content, provision and uptake of vocational training. This design of provision and accreditation would increase not simply the 'portability' of qualifications, but the mobility of labour. In practice, an emphasis on equity and access has taken a back seat to concerns with system expansion (Smith, 1997).

The credentials obtained from participation in industry training will only have value if farmers view it as a meaningful basis for distinguishing more from less productive labour. These questions have not been pursued in any research or programme evaluation thus far conducted in New Zealand, either by government agencies directly, or by government subsidised entities such as the industry training organisations. In addition, decision making and data collection processes provide only cursory checks on the economic value of the training provided. Whether training is government, or industry, organised and provided, and whether it is publicly or privately accredited, it is these questions which must be addressed if the economic value of the agriculture training system is to be assessed and if the form of training provided is to be responsive to the needs of agriculture in New Zealand.

The radical restructuring of the relationship between agriculture and the state has been accompanied in the 1990s by the restructuring of agricultural industry training involving Skill New Zealand. The market-oriented rhetoric and the associated organisational structures established in keeping with Skill New Zealand's strategy seemed to set the course towards greater equality and greater efficiency in the agricultural labor market. However, it is by no means evident that they have had any effect on the agricultural labour market, much less those intended.

Notes

1   Growth in 1997 includes the incorporation of the Wool ITO and trainees previously registered under Wool ITO Training into the agITO figures, rather than this level of growth in new areas of training or a new catchment of participants in training from the agricultural labour force (Cutler, 1992).
2   The notion that wages provide an adequate indicator of productivity differentials is problematic both in general and in relation to agriculture specifically. Because of a variety of features of labour markets, individuals and social characteristics of work, wages are a very imperfect measure both of productivity and of the value of education in the job. Non-formal work (of particular importance in the agricultural sector), internal labour markets and segmented labour markets affect the degree to which wages offer a meaningful indicator of productivity.

References

Addison, J. and Siebert, W. (1994), 'Vocational Training and the European Community', *Oxford Economic Papers*, Vol. 46, pp. 696-724.
Agriculture Industry Training Organisation (1996), *Annual Report*:Wellington.
Anderson, D. (1993), *The Division of Domestic Household Labour on Family Farms*, MA Thesis, Department of Geography, University of Otago: Dunedin.
Benediktsson, K., Manning, S., Moran, W. and Anderson, G. (1990), *Participation in Raglan County Farm, Households in the Labour Force*, Occasional Publication 27, Department of Geography, University of Auckland: Auckland.
Caillods, F. (1994), 'Converging Trends Amidst Diversity in Vocational Training Systems', *International Labour Review*, Vol. 133, No. 2, pp. 241-257
Campbell, H. (1994), *Regulation and Crisis in New Zealand Agriculture: The Case of Ashburton County*, 1984-1992, unpublished PhD Thesis, Charles Sturt University: Wagga Wagga.
Carnoy, M. (ed.) (1995), *Encyclopedia of the Economics of Education*, Elsevier Science: Oxford.
Carnoy, M. (1996a), 'Education and Racial Inequality: The Human Capital Explanation Revisited', *Economics of Education Review*, Vol. 15, No. 3, pp. 259-272.
Carnoy, M. (1996b), 'Race, Gender and the Role of Education in Earnings Inequality: An Introduction', *Economics of Education Review*, Vol. 15, No. 3, pp. 207-211.
Carnoy, M. and Gong, W. (1996), 'Women and Minority Gains in a Rapidly Changing Local Labour Market: The San Francisco Bay Area in the 1980s', *Economics of Education Review*, Vol. 15, No. 3, pp. 273-287.
Cloke, P. (1989), 'State Deregulation and New Zealand's Agricultural Sector', *Sociologia Ruralis*, Vol. 29, pp. 34-48.
Cloke, P. and Le Heron, R. (1994), 'Agricultural Deregulation: The Case of New Zealand', in Lowe, P., Marsden, T. and Whatmore, S. (eds), *Regulating Agriculture*, David Fulton: London.
Codd, J. (1990), 'Policy Documents and the Official Discourse of the State', in Codd, J., Harker, R., and Nash, R. (eds), *Political Issues in New Zealand Education*, Dunmore Press: Palmerston North, pp. 133-149.
Codd, J., Harker, R. and Nash, R. (eds) (1990), *Political Issues in New Zealand Education*, Dunmore Press: Palmerston North.
Cohn, E. (1979), *The Economics of Education*, Ballinger Publishing Company: Cambridge, Massachusetts.

Cutler, T. (1992), 'Vocational Training and British Economic Performance: A Further Instalment of the British Labour Problem?', *Work, Employment and Society*, Vol. 6, No. 2, pp. 161-183.

Dalziel, P. and Lattimore, R. (1996), *The New Zealand Macroeconomy: A Briefing on the Reforms*, Oxford University Press: Melbourne.

DeFreitas, G. (1995), 'Segmented Labour Markets and Education', in Carnoy, M. (ed.), *Encyclopedia of the Economics of Education*, Elsevier Science Inc: Oxford, pp. 39-44.

Dickson, J. (1996), personal communication, Education Training and Support Agency, July.

Doeringer, P. (1995), 'Internal Labour Markets and Education', in Carnoy, M. (ed.), *Encyclopedia of the Economics of Education*, Elsevier Science Inc: Oxford, pp. 28-33.

Durham, K. (1996), personal communication, public relations, agITO, April.

Education and Training Support Agency (1993a), *Annual Report 1992-1993*: Wellington.

Education and Training Support Agency (1993b), *Industry Training Organisations: Responsiveness to People Under-Represented in Training*: Wellington.

Education and Training Support Agency (1994), *Annual Report 1993-1994*: Wellington.

Education and Training Support Agency (1995), *Annual Report 1994-1995*: Wellington.

Education and Training Support Agency (1996a), *Annual Report 1995-1996*: Wellington.

Education and Training Support Agency (1996b), *Funding Schedule for Industry Training Fund*: Wellington.

Education and Training Support Agency (1996c), *Industry Training Fund 1997: Application Guidelines and Forms*: Wellington.

Education and Training Support Agency (1996d), *Industry Training Fund 1997: ITF Operating Guide*: Wellington.

Education and Training Support Agency (1996e), *Skill New Zealand: A Stocktake*: Wellington.

Education and Training Support Agency (1996f), *Training - The Power to Perform*: Wellington.

Education and Training Support Agency (1997a), *Industry Stock Figures at 30 June 1997*: Wellington.

Education and Training Support Agency (1997b), *ITO Trainee Participation Data*, unpublished anayses of in-house data: Wellington.

Education and Training Support Agency and New Zealand National Qualifications Authority (1994), *Taking the Step to Skill New Zealand: A Guide for Employers*: Wellington.

Fairweather, J. (1989), *Some Recent Changes in Rural Society in New Zealand*, Discussion Paper 124, Agribusiness and Economics Research Unit, Lincoln College: Lincoln.

Fairweather, J. (1992), *Agrarian Restructuring in New Zealand*, Discussion Paper 213, Agribusiness and Economics Research Unit, Lincoln College: Lincoln.

Fairweather, J. (1995), *Farm Women and Men's Decisions Regarding Working On or Off Farm*, Agribusiness and Economics Research Unit, Lincoln College: Lincoln.

Higgins, J. (1994), 'Skills and Schemes: The Training Response to Youth Unemployment', in Morrison, P. (ed.), *Labour, Employment and Work in New Zealand*, Wellington, pp. 195-205.

Kearns, P. (1993), *A Review of Research and Development Structures and Practices for Vocational Education Training and Employment in Five OECD Countries*, Australian Vocational Educational Employment and Training Advisory Committee: Canberra.

Keating, N. (1994), 'Family, Gender and Sustainability: Studying the Farm Family', in Bryden, J. (ed.), *Towards Sustainable Rural Communities,* University of Guelph: Guelph.

Keating, N. and Little, H. (1991), *Generations of Farm Families: Transfer of the Family Farm in New Zealand*, Research Report 208, Agribusiness and Economics Research Unit, Lincoln College: Lincoln.

Knoke, D. and Ishio, Y. (1994), 'Occupational Training, Unions, and International Labour Markets', *American Behavioral Scientist*, Vol. 37, No. 7, pp. 992-1016.

Lauder, H. (1991), *The Lauder Report: Tomorrow's Education, Tomorrow's Economy*, New Zealand Post Primary Teachers' Association: Wellington.

Lawrence, G. (1987), *Capitalism and the Countryside: The Rural Crisis in Australia*, Pluto Press: Sydney.

Le Heron, R. (1996), 'Farms, Fisheries and Forests', in Le Heron, R. and Pawson, E. (eds), *Changing Places: New Zealand in the Nineties*, Longman Paul: Auckland, pp. 120-168.

Le Heron, R., Roche, M., Johnston, T. and Bowler, S. (1991), 'Pluriactivity in New Zealand's Agro-commodity Chains', in *Proceedings of the Rural Economy and Society Section of the Sociological Association of Aotearoa*, pp. 41-55.

Liepins, R. and Campbell, H. (1997), *Men and Women as Stakeholders in the Initiation and Implementation of Sustainable Farming Practices: Organic Farming in Canterbury*, Studies in Rural Sustainability, Research Report No. 3, Department of Geography, University of Otago: Dunedin.

Manthei, M., Patterson, J. and Logan, G. (1996), *Māori and Gender Responsiveness Projects*, Agriculture Industry Training Organisation: Wellington.

Min, W. (1995), 'Vocational Education and Productivity', in Carnoy, M. (ed.), *Encyclopedia of the Economics of Education*, Elsevier Science Inc: Oxford, pp. 140-145.

Moock, P. and Addous, H. (1995), 'Education and Agricultural Productivity', in Carnoy, M. (ed.), *Encyclopedia of the Economics of Education*, Elsevier Science Inc: Oxford, pp. 130-139.

National Advisory Council on the Employment of Women (1993), *Report on Women's Access to Industry Training*: Wellington.

New Zealand Employers Federation (1996), *Building Your Business*, New Zealand Employers Federation and New Zealand Qualifications Authority: Wellington.

New Zealand Qualifications Authority (1991), *Designing the Framework*: Wellington.

New Zealand Qualifications Authority (1993), *A Future with Standards: A Guide to the National Qualifications Framework and its Systems*: Wellington.

New Zealand Qualifications Authority (1996a), *Realising the Goals of the National Qualifications Framework*: Wellington.

New Zealand Qualifications Authority (1996b), *Working For Industry. An Introduction to Workplace Assessment and the National Qualifications Framework*: Wellington.

New Zealand Qualifications Authority (1997), *National Qualifications Framework Update*: Wellington.

Olssen, M. and Matthews, K. (eds) (1997), *Education Policy in New Zealand: The 1990s and Beyond*, Dunmore Press: Palmerston North.

Pomeroy, A. (1988), *The Politics of Inequality: Farming Women*, unpublished paper presented at the Conference of Sociological Association of Australia and New Zealand: Canberra.

Purves, A. (1988), 'General Education and the Search for a Common Culture', in Westbury, I. and Purves, A. (eds), *Cultural Literacy and the Idea of General Education*, National Society for the Study of Education: Chicago, pp. 1-8.

Rivers, M. (1992), *The Contribution of Women to the Rural Economy*, MAF Policy Technical Paper 92/4, New Zealand Ministry of Agriculture and Fisheries: Wellington.

Smelt, S. (1995), *Industry Training Organisations*, Education Forum: Wellington.

Smith, J. (1997), Executive Director, Sports, Fitness, Recreation ITO, personal communication, September.

Taylor, C. and Little, H. (1995), *Means of Survival? A Study of Off-farm Employment*, Taylor and Barnes Associates: Christchurch.

Taylor, C., Little, H. and McClintock, W. (1997), *Entrepreneurship in New Zealand Farming: A Study of Farms with Alternative Enterprises*, Technical Paper 97/7, New Zealand Ministry of Agriculture: Wellington.

Willis, R. (1991), *Agricultural Change in New Zealand Since 1984*, Occasional Paper No. 2, Department of Geography, Victoria University: Wellington.

Wilson, O. (1994), 'They Changed the Rules: Farm Family Responses to Agricultural Deregulation in Southland, New Zealand', *New Zealand Geographer*, Vol. 50, No. 1, pp. 3-13.

Winter, M. (1997), 'New Policies and New Skills: Agricultural Change and Technology Transfer', *Sociologia Ruralis*, Vol. 37, No. 3, pp. 363-381.

10 Discourses of community and dairy company amalgamations in Taranaki, New Zealand

CAROLYN MORRIS

'History is not simply something that happens to people, but something they make - within, of course, the very powerful constraints of the system within which they are operating. A practice approach attempts to see this making...' (Ortner, 1984: 159)

Introduction

Events and processes of change in agricultural production and rural life that are written about as operating at international and national levels actually occur in specific localities, and happen to particular groups of people.[1] People experience, understand and seek to influence change (which is often perceived as emanating from outside), through localised idioms and interpretive frameworks, constructed in the context of wider socio-cultural models and theories. A common discourse for understanding rural change, both among academics and farmers, is in terms of impacts on communities. But while change is often experienced as an 'impact' by farmers, they are not passive in the face of change. Instead, they work to resist, ameliorate and/or incorporate change in terms of particular local discourses and practices. In rural areas the discourse and practice[2] of community is a common strategy for managing change. But while community discourse might be widespread, discourses of community are highly particular and specific to localities. So in order to understand the processes of change in a locality, it is necessary to discover what interpretive frameworks local people are using.

Between 1985 and 1989 there occurred a series of dairy factory amalgamations in Coastal Taranaki which resulted in serious social conflict among local dairy farmers. This chapter describes that social conflict and, using the theoretical work of Anthony Cohen (1985), explores in some detail how local farmers explained and understood the course of events through a particular discourse and practice of locality and community, expressed and performed through the symbol of 'the Coast.'

According to Cohen (1985: 12), community is about the marking of similarity and difference, about expressing the distinction between an 'us' and a 'them'. This distinction is asserted by people as and when the need arises. The distinction operates on two levels; it is used to tell outsiders what the community is, and it is used by members of the community to tell themselves who they are; in other words, it is about the construction of their own identity. Community, thus, is a symbol which people use to express their sense of belonging to a locality and/or to a group of people and to express their difference from others. Cohen argues that the boundary of community is where analysis ought to be focused as it 'encapsulates the identity of the community' (Cohen, 1985: 12). As local communities become increasingly integrated into national and global political and economic systems, the structural boundaries that distinguish them break down. However, this does not result in homogeneity and the end of diversity. Rather, boundaries between the community and the outside will be reinforced or reconstituted symbolically; because 'as the structural bases of boundary become blurred, so the symbolic bases are strengthened through flourishes and decorations, aesthetic frills and so forth' (Cohen, 1985: 44). The discourse of community is brought into play when people feel that their community is under attack, and when the possibility arises that the boundary of the community may be breached. People attach their own meanings to symbols and, Cohen points out, if the symbol is a common part of everyday life and discourse, people may not realise that they do not attach the same meanings to the same symbol (Cohen, 1985: 17). Community works as an aggregating rather than an integrating symbol (Cohen, 1985: 20) in that it allows people to express their similarity and commonality without losing their individuality. The symbol of community has different meanings for different people because people experience community in different ways, through family, religion, locality, occupation, age, gender, class, ethnicity, life course and so on (Cohen, 1985: 88). So different people will use the discourse of community in different ways and in different contexts to achieve different ends.

In the case of the dairy factory amalgamations on the Coast, the discourse of community was used by a majority of farmers to resist the threat to local identity posed by the removal of the key symbol of community, the cooperative dairy company.

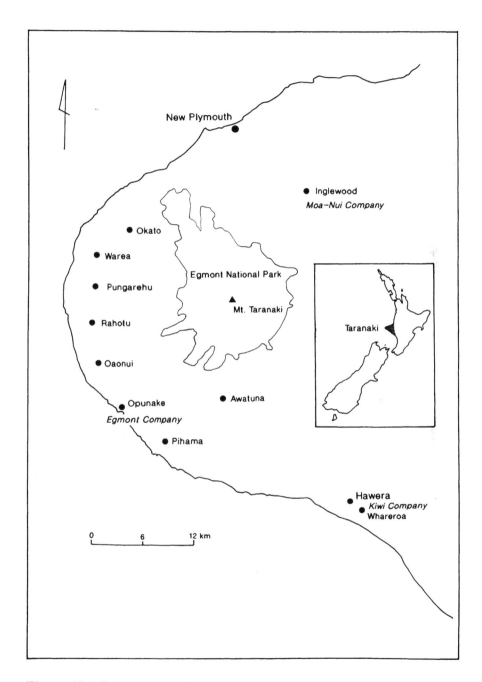

Figure 10.1 Taranaki and Egmont dairy region

Farmers became aware that while they used a common discourse of Coast and community, in practice it meant very different things to different people. In Cohen's terms, they held the symbol in common, but not the content. Farmers were forced to reconsider the meaning of the 'Coast and community' and to find new ways to practice community. The Coast, as the symbol of community, became detached from its referents and new meanings had to be generated. Social conflict became intense because the rupture of the boundary and the challenge to the integrity of community identity came from inside, not outside of the boundary.

The amalgamations of the dairy factories

The dairy industry in New Zealand is a co-operative enterprise, with supplying farmers owning their local manufacturing company, and through these companies the assets of the New Zealand Dairy Board, which is the sole exporter of dairy products from New Zealand.

The history of the dairy industry on the Coast is one of expansion as settlements grew in size and number.[3] In the post-War War Two period, there were 13 companies with 22 factories, but the advent of milk tankers in the 1950s and 1960s meant that farmers no longer had to transport their milk to the factory, and there followed a process of amalgamation and consolidation of factories along the main Coastal road. During the 1970s, there was a further series of amalgamations so that by 1985 there was only one company left on the Coast, the Egmont Company at Opunake.

At the beginning of the 1989/1990 season, around 50 farmers, the majority of whom lived in the Okato district, left the Egmont company and shifted their milk supply to the Kiwi company at Hawera. The proximate cause of this 'defection', as it was widely described by farmers, was the 1988 Commerce Act. Under this Act, supply area agreements between dairy companies were judged to be monopolistic and therefore illegal. Until then, Taranaki dairy companies had boundary agreements and would not take suppliers from each other's areas, which meant that farmers could not change companies. This event caused a great deal of conflict and anguish among farmers on the Coast and was cited as the first in a series of events that led to the eventual amalgamation of the Egmont company with Moa-Nui in 1989, and the end of dairy manufacturing on the Coast.

Farmers explained the series of amalgamations in several ways. The causes they pointed to operated at a variety of levels - global, national and local - and included, for example, the UK joining the EEC in 1974; the long term decline and fluctuations in global commodity prices for dairy products; the system by which the Dairy Board paid manufacturing companies for their milk (which pushed companies to increase in size to achieve economies of scale); legislation such as the Commerce Act of 1988; declining payouts from particular local companies; and the local politics of community identity. Dairy company amalgamations in general were explained as the result of long-term macro-economic changes that were beyond local control, and the motives of particular companies in seeking amalgamation were considered to be primarily economic. However, it was local factors that were considered to account for the seriousness of the 'troubles' that arose over the amalgamations.

'The Troubles'

In 1992, the Moa-Nui Cooperative Dairy Company sent a newsletter to all of its suppliers which, in an article entitled 'End of Taranaki Land Wars', contained the following statement:

> The rivalry between Taranaki's dairy companies in recent years has been almost as distressing as the land wars of last century. Family friendships that went back generations have been sundered by changes in affiliation to one dairy company or another. The arguments have gone far beyond payout; company philosophy, personality of directors, location of factories - even design of logos and the new name of amalgamated companies - have brought neighbours close to blows, caused fights in school playgrounds and ruined friendships (*Moa-Nui Company Newsletter*, February 1992).

The war that this quote refers to is the series of battles over dairy company amalgamations that occurred in Coastal Taranaki between 1985 and 1989.[4] The 'trouble' that arose over this amalgamation was also compared by several informants to the 'troubles' in Northern Ireland. Obviously this 'trouble' was nowhere near as serious as the land wars of the 1860s or the troubles in Northern Ireland. It amounted to one or two fights at local pubs, some fighting among children at school, a few stones being thrown at tankers, and the destruction of some factory supply number signs.[5] The trouble largely manifested itself in talk. Talk now had 'edges', or was 'frosty', people shouted and argued, or stopped speaking to each other altogether. In the end, public discussions about dairy factory business, a perennial topic of conversation amongst farmers, were taboo. (Upon reading transcripts of their interviews, several farmers commented that the trouble looked rather insignificant when written down and yet it felt very serious to them and they considered it to be highly significant).

As with social scientists, ordinary people have a range of theories available to them with which they explain and understand situations and events. To understand the amalgamation troubles, it is necessary to understand how local farmers conceptualised 'the Coast', for the structure of 'the Coast' was proffered as a key explanation for much of the trouble over the amalgamation. 'The Coast' and 'community' are potent and closely-linked symbols in local discourse, considered to be descriptive of reality and to have powerful explanatory and persuasive ability. What I aim to show in this chapter is what happened on the Coast when the discourse of 'community and Coast' could no longer account for the reality it was intended to describe.

Local discourses of community

The people who live on the Coast described themselves as having different interests from, and being significantly different to, people in the rest of Taranaki. They considered themselves to be a community. Informants talked about the Coast as a clear and distinct entity, as a separate and definable area that was significantly different from surrounding districts. The key terms people used in describing themselves and the Coast as distinctive were isolation; separation; marginalisation; independence; self-reliance; initiative and strong community spirit.

The perceived marginalisation and isolation of the Coast were seen to have fostered in the people a sense of independence, self-reliance and 'strong community spirit'. To quote one informant:

> The difference between the Coast and the rest of Taranaki is that around the Coast people have more of a community sense, a sense of belonging. It's because of being cut off by the mountain and the sea. In other areas like Stratford [around the other side of the mountain] they don't have that same sense of belonging.

Informants also linked this strong sense of community to the dairy industry on the Coast. 'Farming is the core of the community', one farmer said. The dairy industry and dairy farming were perceived to be a major source of unity and commonality among the people of the Coast, and the dairy companies, people said, were at the core of communities. To quote another informant:

> The local dairy companies have been the centre of each community in Coastal Taranaki, you were proud of it, you looked after it. One of the strengths of dairy farming communities was the feeling of being at one with each other, and the thing that was there was the dairy factory. That made you be at one with each other, and made you be part of that community, it was something special.

However, the image of the Coast as a unified and united community is only partial. When the occasion demanded, people would agree to a certain depiction of themselves as 'the Coast', but they said that in reality the Coast was far more complex and far less cohesive than the picture they presented to outsiders.

The Coast is considered to be divided into 'ends', with the dividing line located between Warea and Pungarehu. This division is expressed in terms of 'different centres of interest', with people from the Okato end looking towards New Plymouth, and with Opunake as the focus for the southern end (e.g. for farm supplies). This division is further marked by Warea and Pungarehu being within in different telephone areas and by the fact that this also serves as the boundary of the catchment areas for the two secondary schools on the Coast, located at Opunake and Okato.

Further, the Coast is divided into individual villages or districts, each with its own character and interests. As one informant suggested, identification with these villages or districts is based on the location of the original dairy factories:

> So you get people saying they're from Oaonui and Rahotu and Kahui Road, but not from Wiremu Road or Namu Road. This is because the dairy company was the focus or the core around which the villages were established. So people say they're from the dairy company they identify with.

The boundaries drawn between the villages follow the old dairy company supply boundaries, and also coincide with primary school catchment areas. The villages are considered to have different and sometimes conflicting interests and to have different characteristics (explained, for example, as the result of size, class, religion and so on), so that some villages are said to be 'stronger communities' than others.

The community and the dairy factory

Until the late seventies or early eighties, each village had a co-operative dairy company owned and directed by resident farmer/shareholders. The dairy factory was at the centre of each of the communities, and to many farmers it symbolised what it meant to be from that village, embodying within it key concepts of Coastal identity: a strong spirit of co-operation, loyalty, and cohesion. The fact that the dairy industry is a co-operative industry means that many of the values and behaviours associated with and essential for the economic success of the co-operative and its members, (ideas like co-operation, working together, loyalty to the company, democratic decision making, the primacy of the group over the individual), are very similar to ideas associated with what is considered to be community-minded behaviour, and with the behaviour and values that are considered locally to be particularly and specifically Coastal. The dairy factory was the symbol *par excellence* of community on the Coast.

When in the 1950s and 1960s, the factories on the side roads amalgamated with the companies on the main roads, there was little disruption as farmers were only shifting their supply to their local village. Already their children went to school there, they did their shopping there and had the focus of their social life there - they went to the pub and the stock sales and Women's Division and so on. There was little trouble over these amalgamations as the closure of the factory did not mean the end of the company and so did not threaten the community.

However, later changes in the dairy industry had a more significant impact on the relationship between factory and community. When the village dairy company amalgamated and the factory closed, it could no longer function as the centre of the community. The loss of each village dairy factory was equated with the loss of village 'identity'. As one farmer said: 'So the Okato company became part of Egmont and we lost our identity'. With that loss of identification between factory and community came a loss of a sense of loyalty to the company:

> The centre of the company is no longer on the Coast, and the original loyalties that used to be existing in your little focal point are no longer there.

So the amalgamation of the village companies into the Egmont company in Opunake from 1976 onward caused much more anguish and argument in the villages. People were afraid that the loss of their company would mean the loss of their identity as a village, and the end of their community. With the loss of the local dairy factory, a central component in the symbolic map of community in the Coastal villages was gone. Previously the economic and the social had coincided in the village, but now the economic base of the community had been removed. This structural change necessitated a reconsideration of the village as a community. With the dairy factory gone, both a structural base and a key symbolic component of community were removed, so the community, if it were to survive as a community, had to be 'built on' other things. People adjusted to this change. Opunake, where the dairy company was now situated, was at least part of the Coast, part of the same wider community. But people did not feel quite the same sense of attachment or loyalty to the Egmont Company as they had done to the local village company.

When the Egmont company amalgamated with Moa-Nui in 1989 the economic base of the Coast was removed from local control, and the most potent local symbol of community was taken away. Inglewood, where Moa-Nui was sited, was around 'the other side of the mountain', in Coastal terms a very different and distant place. This event coincided fairly closely in time with local government restructuring and the dissolution of the Egmont County, which had more or less coincided with the Egmont Dairy Company supply area. In the space of 18 months, what were considered to be two of the three major structural forms that defined the Coast were gone.[6] The co-operative dairy company was no longer available as a focus of community identity and people feared that this would be the end of their community.

According to many farmers on the Coast, the Egmont/Moa-Nui amalgamation occurred because of the actions of those 50 farmers (or 'traitors') from Okato who, as noted earlier, had broken away from the Egmont Company and taken their milk to the Kiwi company. From this perspective, the community was not only under attack from outside forces, but from within as well, by people who were members of that community. According to Cohen (1985: 12), community is essentially about the marking of similarity and difference: people within a community are considered to be more like each other than they are like people from the outside. As such, they can be expected to behave in ways which demonstrate, and reinforce, this distinction. The actions of the farmers who went to Kiwi were outside the range of difference containable by the concept of 'the Coast'. Their economic interests were now allied with people from a different place, from Hawera. The boundary between 'them' and 'us' had changed, and 'us' could no longer be expressed by the concept of 'the Coast'.

There was reportedly no surprise farmers from Okato who broke away and took their milk to Kiwi. Okato and Opunake, at opposite ends of the Coast, mark the limits of community. People identify more strongly with their village than they do with the Coast and Coastal identity was not strong enough to influence farmers from Okato to behave in a community-like manner with the people from Opunake.

The people on the Coast were faced with a major contradiction. The group from Okato had done something that many considered a person from the Coast would not do, and yet these people clearly were from the Coast. As a result of the amalgamation, 'the Coast' became a problematic concept (perhaps for the first time for many people). The discourse of 'Coast and community' proved inadequate as a description of reality and as an explanation of behaviour. What now did it mean to come from the Coast? In what ways were people from the Coast different from people elsewhere? The previous construction of the Coast could no longer be held. The dairy factory, *the* symbol of Coastal identity was gone, and people were doing things that by definition people from the Coast would not do.

Eventually the troubles died down. Dairy factory business, once a perennial topic of conversation, was taboo in public for a while. Now, many simply find the topic uninteresting. Farmers on the Coast no longer express a sense of identification with and loyalty to the company. Instead, there is cynicism about the company which, many farmers say, is no longer working in their interests even though they are the owners. As a result, it may be difficult for the company to mobilise farmers in the name of the company, which may have implications for the future of the cooperative nature of the industry.

Conclusion

The 'troubles' caused by the amalgamation of the dairy factories presented a radical challenge to the concept of community as understood and used by Coastal farmers, but it did not result in the demise of community discourse. Instead there has been a shift in their understanding of what community is, and community and economics have been separated. On the Coast economic relationships, in the form of the cooperative dairy company, were considered to be part of the community, even the basis of the community, but with the end of the Coastal cooperative dairy companies, this is no longer so. Discussions about the dairy industry are now conducted in economic, not community, terms. Prior to the amalgamation,s economic relationships among farmers on the Coast were cooperative, not competitive, and the discourse of community was used by many farmers on the Coast to resist the establishment of competitive economic relationships among them. The pressure for the establishment of such relationships was generally regarded as emanating from *outside* the community and as being a threat to community, and community was considered to be a site of resistance to such forces. It was the fact that these competitive relationships were actually created by people from *within* the community that led to the trouble over the amalgamations, and which were regarded so seriously as to be compared to the troubles in Northern Ireland and the Land Wars of the last century.

Community studies have fallen out of fashion in the social sciences for several reasons, including the inability of researchers to locate communities empirically, and the ineffectiveness of 'community' as an explanatory tool. However, as the amalgamation story shows, Coastal farmers insist that they do live in real, empirically locatable communities, and for them community has considerable explanatory power. A discourse/practice approach to the study of community, and the use of Cohen's (1988) analytic focus on the symbolic nature of community as expressed in boundaries, allows the researcher to overcome some of the problems associated with more structural approaches. Particular people and groups of people use the discourse of community in particular contexts to achieve particular goals, and close examination of this discourse can contribute much to an understanding of processes of social and economic change in agriculture and rural life.

Notes

1   This paper draws on my MA thesis (Morris, 1993). It was based on interviews with 33 farmers in Coastal Taranaki, carried out in 1990/1991.
2   I am using Abu-Lughod's (1991: 147-48) definitions of the terms discourse and practice which, drawing on the work of Bourdieu (1977) and Ortner (1984), point to 'the social uses by individuals of verbal resources' and people's 'strategies, interests and improvisations', and which 'refuse[s] the distinction between ideas and practices, or text and the world' (Abu-Lughod, 1991: 147-148). A discourse/practice approach recognises the different and contradictory understandings and goals that exist within groups, and the resulting real outcomes of these.
3   The first cooperative dairy company on the Coast was established at Opunake in 1885.
4   The original Taranaki Land Wars were fought between Maori and Pakeha in the 1860s and resulted in the large-scale confiscation of Maori land
5   Dairy farmers have small metal signs at the road end of tracks to cowsheds with their company supply number on it.

6   The third major force was the geography of the Coast, and the location between the mountain and the sea.

References

Abu-Lughod, L. (1991), 'Writing Against Culture', in Fox, R. (ed), *Recapturing Anthropology: Working in the Present*, School of American Research Press: Santa Fe.

Bourdieu, P. (1977), *Outline of a Theory of Practice*, Cambridge University Press.

Cohen, A. (1985), *The Symbolic Construction of Community*, Ellis Horwood: London.

Morris, C. (1993), *Understanding the Egmont Amalgamation: Community in Taranaki and Anthropology*, M.A. thesis, Department of Anthropology, University of Auckland: Auckland.

Ortner, S. (1984), 'Theory in Anthropology since the Sixties', *Comparative Studies in Society and History*, Vol. 26, pp. 126-166.

Williams, R. (1983), *Keywords: A Vocabulary of Culture and Society*, Fontana: London.

PART III
AGRICULTURAL POLICY AND POLITICS IN AUSTRALIA AND NEW ZEALAND

11 Markets as politics: The case of the meat export industry of New Zealand

BRUCE MACDONALD CURTIS

Introduction

This chapter extends the observations by Moran *et al.* (1996a) that producer marketing boards like New Zealand's Meat Board have operated as a grand averaging scheme for farmers. It provides an institutional explanation in terms of the interconnection of state and markets (Curtis, 1996). This is attempted by presenting 'markets as politics' (Fligstein, 1996). Fligstein has outlined how markets are the fluid product of struggles between actors to create stable worlds and lasting solutions to the problems of competition and market development. States are inevitably drawn into these political struggles to organise markets. In this sense, all forms of stability in markets are state-sponsored and at the same time, are subject to transformation from exogenous forces, including 'invasion, economic crisis, or political intervention from states' (Fligstein, 1996: 669).

The play-off between state attempts to organise stable markets and exogenous forces which tend to de-stabilise is nowhere more apparent than in the case of the meat export industry. What must be added to Fligstein's propositions, so as to properly represent the case of the meat export industry in New Zealand, is the willingness of the state to systematically favour one set of actors (farmers) over another (meat firms) (Curtis, 1992; 1996). This caveat to Fligstein's propositions about 'state-building as market-building' means this discussion of markets as politics in the sphere of agriculture might also be of use to those interested in accounts of settler capitalism (Boreham *et al.*, 1989; Denoon, 1983; McMichael, 1984), and Fordism and food regimes (Britton and Le Heron, 1986; Friedmann and McMichael, 1989; McMichael and Myhre, 1991; Moran *et al.*, 1996b; Roche, 1992).

My argument is that the Meat Board, a statutory body, has organised markets against meat firms in the collective interests of farmers. However, the most important markets, and the divisions between farmers and meat firms, are now far more problematic than when the Meat Board was established in 1923. This transformation in markets recasts the politics of intervention, especially because the ensuing re-organisation of markets is expressed mainly through the rise of farmers' co-operative meat firms.

The Meat Producers' Board and the consignment system

The Meat Board (formerly known as the Meat Producers' Board) was established by the Meat Export Control Act in 1923 (Hayward, 1972). The context of this Act was a dramatic slump in the prices for meat and other agricultural goods exported to all-important markets in Britain, and the subsequent mobilisation of farmers in New Zealand demanding relief and protection.

The Meat Producers' Board (the Board) was from the outset the farmer's statutory body. The Board was comprised of a majority of representatives elected by farmers. The composition of the Board included five 'representatives of the producers of meat for export', two 'representatives of the New Zealand Government' and one 'representative of persons for the time being engaged in business as stock and station agents' (Meat Export Control Act, 1923). However, the delegate proposed by the stock and stationers was an outspoken critic of the Board and was rejected by the Minister of Agriculture. Indeed, the position for a representative of the stock and stationers remained unfilled until it was abolished in 1956. Arguably, the legislation which created the Meat Producers' Board in 1923 (and the Dairy Board in 1924) epitomised the antipathy farmers commonly show for 'middlemen' (Stinchombe, 1961). It certainly demonstrated the strength of farming groups in New Zealand in 1923 (Font, 1990).

From the perspective of farmers there was a dire need for the Board. By 1923 there was considerable evidence to suggest that the largest of the meat firms, including Vesteys, Borthwicks and Swifts, had deliberately crashed prices for stock and meat so as to extend their sway over the international trade in meat. These were transnational companies (TNCs) that exported significant tonnages of frozen and chilled meat to Britain, not just from New Zealand but also from Australia, Argentina, Uruguay and Brazil. The strategy of crashing prices had been employed previously by the TNCs to bankrupt and buy up their smaller competitors and farms, but had been stalled by the outbreak of war in 1914. The end of the war, and of the war commandeer in

1920, provided the opportunity for the TNCs to resume their tested strategy for growth (Chandler, 1978; 1990).

Such growth, in the form of vertical integration by the TNCs, posed a serious problem for farmers. In short, vertical integration would eliminate the variety of contracting arrangements farmers could use to sell stock to meat firms or to sell meat on their 'own account'. This contracting organisation of markets was called the 'consignment system'. Consignment fostered diversity and specialisation in the spheres of processing and marketing. It secured small processing firms without forward links into marketing as well as small marketing firms without backward links into processing. These specialised firms acted as the agents of farmers (and others) and gained their revenues from commissions and charges (Critchell and Raymond, 1912). If unchecked, vertical integration by the TNCs would displace the consignment system and cause farmers' incomes to plummet further (Smith, 1969).

Stable terminal markets

In response to the activities of the TNCs, the Board acted to secure the collective interests of farmers by making strategic interventions in the organisation of local and global markets. These interventions allowed an extension of the benefits of consignment within the context of the growth of the TNCs. They were realised across commodity chains (Gereffi *et al.*, 1994) where the terminal markets for meat were located more or less exclusively in Britain and were secured through the paucity of the British 'home kill'. For as long as Britain remained the prime destination for meat-exports the Board's efforts at organisation centred on trade-offs with the TNCs. The interventions of the Board meant that the TNCs (operating in New Zealand) were required to use vertical integration as well as a variety of contracting arrangements with farmers and small firms to secure stock and processing facilities.

Control of freight

At its first meeting in 1923 the Board took over all aspects relating to the negotiation of freight rates and the scheduling of shipping. Previously these arrangements had been negotiated between the owners of the cargoes of meat-exports (or their agents) and the shipping lines. The Board's intervention meant that shipping lines no longer made deals with farmers or firms with a direct interest in cargoes. The Board acted as the single buyer of shipping. It negotiated a single, averaged freight rate for all cargoes and arranged schedules for the delivery of cargoes to ports around New Zealand and their loading on board.

This intervention had two dimensions. By forcing the powerful shipping lines to negotiate with a single buyer it dampened increases in freight rates. More significantly, it meant that farmers and their agents, small firms and the TNCs had equal access (priority in queuing) to shipping and paid the same rate for freight regardless of the tonnages they shipped. This form of control meant that the shipping lines and the TNCs could not collude to squeeze out the smaller players. This was very significant for farmers as at least one of the TNCs - Vesteys - owned its own shipping line (Knightley, 1981).

Licensing of meat-export slaughterhouses

The most clear-cut check to the ambitions of the TNCs took the form of the licensing of meat export slaughterhouses. This was undertaken at first by informal mechanisms. The Board had inherited a comprehensive system of inspection of processing facilities relating to hygiene standards (the Slaughtering and Inspection Act of 1900). Prior to the establishment of the Board these checks on hygiene were administered by local authorities and the Department of Agriculture. They were intended to ensure the safety of residents living near slaughterhouses and, to a lesser extent, of consumers in Britain. The checks meant that only output from slaughterhouses with a satisfactory standard of hygiene was granted a meat export license. Such a license was required for every cargo exported from New Zealand. However, the Board transformed the supervision of hygiene standards into the effective licensing of the ownership and use of processing facilities. This was achieved through a liberal interpretation of the Slaughtering and Inspection Act (1900) which the Board was now charged to administer. For example, the Board made it known to the North American firm Armour that should it succeed in acquiring processing facilities in New Zealand, any output from its plants would be denied a meat export license. This threat ended the attempt by Armour to buy processing facilities in New Zealand. By the same mechanisms the Board in 1929 placed restrictions on the output from slaughterhouses already owned by Vesteys and Borthwicks. These restrictions were not lifted until 1952 (Harrison, 1963; Knightley, 1981).

This form of control was given statutory authority by the Meat Act (1934) which gave the Board full powers to license meat export slaughterhouses as well as the output of these facilities. Through the licensing of meat export slaughterhouses the Board directly controlled the entry of new firms to the industry; takeovers or mergers by existing players; and the level of throughput at each slaughterhouse. By this form of control, the Board ensured the survival of many small plants and firms. At the same time, it secured a disposition of plants and firms so that nearly every farming district in New Zealnd enjoyed intense competition for stock. That is, the Board enforced overlapping catchments for stock on plants and firms to the great benefit of farmers acting as suppliers.

The policy of the 'open door'

The Board enforced the right of farmers, acting as individuals or in groups, to use the processing facilities of meat export slaughterhouses. After 1934 the licensees of meat export slaughterhouses were required by law to accept for slaughter all stock offered by farmers that were intended for export, and to handle that stock on the farmers behalf. Licensees (meat firms, large and small) were required to process and grade stock intended for consignment or for cooperative marketing ventures. This exceptional demand on the holders of export licenses was called the policy of the 'open door'.

Through the open door, farmers who chose not to sell stock to meat firms (at the farm-gate), had ready access to processing facilities needed to export meat on their own account. As a result, the open door allowed farmers an alternative avenue for selling whenever they felt that meat firms were offering too low a price for their stock. The open door was also used by a number of farmer-owned firms to begin their marketing operations without the capital outlays required to build or buy processing plants.

Unstable terminal markets

The initial interventions by the Board had three dimensions: to check the advance of the TNCs; to preserve vestiges of the consignment system; and to support prices for stock in local markets. These interventions reflected efforts by the Board to best position farmers in commodity chains where issues of active marketing (Cornish, 1995) were relatively unimportant (thanks to the composition of markets in Britain). However, Britain was lost as a secure terminal market over several decades. This began in 1954 with the end of the World War Two era war commandeer and was most clearly symbolised by Britain's entry into the European Community in 1972. This transformation of markets in Britain impacted on all actors in the industry and on the nature of interventions by the Board. As a result, the Board was forced to become involved in new areas of intervention. These new interventions overlaid, but did not displace, the existing efforts of the Board to maximise the income of farmers.

The schedule and price support

The Board introduced two very different versions of price support in 1958. Both versions operated to enhance the prices farmers received for stock. Firstly, the Board ensured that farmers could more readily play-off competitors for stock. To do so the Board published a schedule for every farming district in the country which outlined the prices per kilogram for different classes of stock offered by firms in the area. These schedules were developed through negotiation between agents of the Board and the firms. Not only did the schedules provide a baseline for negotiation between farmers and firms, but they also required that firms offer prices for all classes of stock at all times of the year. This aspect of the schedule facilitated the traditional (that is, low cost) farming practices of sheep and beef farmers. Secondly, the Board also moved into direct price support for farmers. Its minimum price scheme was used intermittently to augment the returns to farmers for specified classes of stock. The minimum price scheme was financed from the levies paid by farmers to the Board and arguably represented a form of price smoothing.

Of far greater consequence was the supplementary minimum price (SMPs) scheme established by the National Government in 1975. The SMPs were funded by the state and were administered by the Board. The SMPs operated fairly continuously until their abolition in the mid 1980s and in doing so they contributed to a very significant increase in farm production.

Managing the exit of the TNCs

Throughout the first decades of its operation, the Board organised markets against the interests of the TNCs. Nevertheless, the Board did not attempt to eliminate or marginalise firms like Vesteys, Borthwicks and Swifts. The TNCs owned the bulk of the refrigerated infrastructure which allowed the connection of markets in New Zealand and Britain. As such the TNCs constituted a threat, but at the same time they were indispensable to farmers and the Board. As long as prices for stock remained high in local markets and farmers were able to use alternative (own account) avenues for selling onto the markets in Britain, then the Board was relatively unconcerned about the dominance of the TNCs. Within this organisation of markets Vesteys, Borthwicks and Swifts were free to export the bulk of meat from New Zealand. In this sense, the Board cannot be said to have adopted a nationalistic perspective. Rather, the Board operated to maximise the incomes of farmers by making sure that the deals they struck with the TNCs were favourable. The Board ensured that the global economies of scale enjoyed by the TNCs were not turned against farmers in New Zealand.

However, stagnation in the volumes and prices of meat exports going to Britain served to destabilise the TNCs. The piecemeal exit of the TNCs from New Zealand (and the international trade in meat) began in 1973 with Swifts, continued with the departure of Borthwicks in 1980, and was completed with the demise of Vesteys in 1995. There are a host of reasons for the collapse of the TNCs, including: their reliance on the British market; their investments in wholesaling; the rise of supermarket chains; the growth of British production; narrowing margins; and mismanagement. There is not room to explore issues of causality here, but it is important to note that the piecemeal exit of the TNCs created major problems for the Board and farmers in the organisation of markets.

The Board attempted to manage the exit of the TNCs through the fostering of a large New Zealand company, Waitaki International, in their place. Waitaki International grew through the takeover of Swifts, Borthwicks and other firms to become the largest meat firm in the world. Unfortunately for the Board, Waitaki International collapsed in 1989. In large part this failure resulted from the Board's focus on positioning farmers in local markets for stock. While the Board allowed Waitaki International to buy up its competitors, it prevented the firm from closing excess processing facilities or from imposing new, unfavourable, deals on farmers.

Continuities and discontinuities

The portrayal of the events in the meat export industry and of the strategies of the Board has emphasised an argument drawn from accounts of the dismemberment of the welfare state. This emphasises the retrenchment of the state in all its forms from social and economic life. In New Zealand this approach is associated with discussions of 'restructuring' (Britton *et al.*, 1992; Cloke, 1989; Lawrence *et al.*, 1992; Nolan, 1994; Perry, 1992). There is compelling *prima facie* evidence found in such accounts of the meat-export industry (Hartley, 1989, Le Heron, 1988; Savage, 1990). For example, all the conditions imposed by the Board have been abolished or relaxed: the licensing of meat-export slaughterhouses and the formal policy of the open door were abolished in 1981; the control of shipping was extended to a meat industry freight council (including the Board and firms) in the mid 1980s; the SMPs were abandoned in 1985; Waitaki International failed in 1989; and the Board ceased publication of its schedule in 1995. Further, the composition and name of the Board was changed in 1996 to include representatives from meat firms, and the current status of the reconstituted Meat Board is under constant review.

Clearly, if the 1980s and 1990s are compared with the 1920s to 1970s, then major discontinuities emerge in terms of the willingness of the state to underwrite the Board and the character of its interventions. However, there are also continuities which emerge only when the meat export industry is conceptualised as a dynamic network. Accounts of restructuring mesh neatly with the broader accounts of the crisis of Fordism and both present 'before and after' models of regulation that are divided by a singular, albeit multifaceted, crisis. Unfortunately, the notion of crisis obliterates the character and motivations of social action, where all solutions are partial, where the struggle for stability is constant, and where crises are normal (Cloke, 1996). From this perspective - of markets as politics - the current interventions of the Board contain continuities as well as discontinuities. In essence, what has changed are the circumstances in which the actors find themselves, rather than their motivation to create stable worlds.

Export licensing

Many of the mechanisms by which the Board could previously control firms have been abolished. These include the licensing of processing plants, the promotion of the open door and the unchallenged control of all aspects of shipping. However, all firms exporting from New Zealand still require export licenses which are granted through the Board. Further, the Board can designate markets as being of special interest and thereafter require firms wishing to sell in these markets to adhere to any guidelines it imposes. As was the case prior to 1934, when the Board used one set of statutory powers (the surveillance of hygiene) to achieve other ends (the de facto licensing of processing), its influence in the sphere of marketing meant that the Board could not be marginalised by any interests wishing to engage in meat export.

Fostering the co-operatives

Currently three farmers cooperatives control roughly two-thirds of the industry. This is a new and decidedly ambiguous situation for the Board and for farmers. For most of their history, farmers' cooperatives were a marginal organisation form for doing business and for securing the interests of farmers. On the one hand, the Board made deals with the TNCs. On the other hand, farmers preferred to sell stock to meat firms, or to sell meat on 'own account'. Indeed, the growth of the farmers' cooperatives must be understood as a response favoured by the Board to fill the void left by the demise of Waitaki International (which was itself favoured by the Board to fill the void left by the TNCs).

Farmers now own the processing and marketing arms of the industry, activities which the Board had for so long served to check on the farmers' behalf. In this respect, farmers can be said to have conflicting interests. Acting as suppliers of stock, farmers wish to maximise their returns in local markets. Acting as shareholders, farmers wish to maximise their returns from investment in their cooperative.

Conclusions

Given the enhanced ownership by farmers of meat firms, the historical willingness of the state/the Board systematically to favour one set of actors (farmers) over another (meat firms) is made problematic. The new, ambiguous position of farmers and the Board is expressed in part through a reworking of the composition of the Board. For much of its history, the Board was dominated by the representatives of farmers and the interests of meat firms were excluded. However, in 1996 membership of the Board was broadened to include four representatives of firms in the industry. This transformation of the Board from a sectoral body (Bartley, 1987) to an industry body could be construed as proof positive for those emphasising the discontinuities of governance. However, the majority of these new representatives come from the farmers' cooperatives which were enlarged in the wake of the collapse of Waitaki International. As such, continuities of governance are also evident in that farmers continue to exercise control in the meat-export industry and that the Board remains the key site for the exercise of this control.

References

Bartley, C. (1987), *The Accountability of the New Zealand Meat Producers' Board to Farmers from 1922-1985*, Unpublished Masters Thesis, University of Canterbury: Christchurch.

Boreham, P., Clegg, S., Emmison, J., Marks, G. and Watson, J. (1989), 'Semi-Peripheries or Particular Pathways: The Case of Australia, New Zealand and Canada as Class Formation', *International Sociology*, Vol. 4, No. 1, pp. 67-90.

Britton, S. and Le Heron, R. (1986), 'Regimes and Restructuring in New Zealand: Issues and Questions in the 1980s', *New Zealand Geographer*, Vol. 43, No. 3, pp. 129-139.

Britton, S., Le Heron, R. and Pawson, E. (1992), *Changing Places in New Zealand: A Geography of Restructuring*, New Zealand Geographical Society: Christchurch.

Chandler, A. (1978), *The Visible Hand: The Managerial Revolution in American Business*, Belknap: Harvard.

Chandler, A. (1990), *Scale and Scope: The Dynamics of Industrial Capitalism*, Belknap: Harvard.

Cloke, P. (1989), 'State Deregulation and New Zealand's Agriculture Sector', *Sociologia Ruralis*, Vol. 28, No. 1, pp. 34-48.

Cloke, P. (1996), 'Looking Through European Eyes? A Re-evaluation of Agricultural Deregulation in New Zealand', *Sociologia Ruralis*, Vol. 96, No. 3, pp. 307-330.

Cornish, S. (1995), 'Marketing Matters: The Function of Markets and Marketing in the Growth of Firms and Industries', *Progress in Human Geography*, Vol. 19, No. 3, pp. 317-337.

Critchell, J. and Raymond, J. (1912), *A History of the Frozen Meat Trade*, Constable and Co: London.

Curtis, B. (1992), 'Product Markets and Labour Markets: The Paradox of Flexibility in the Export Meat Industry', in Morrison, P. (ed), *Labour Employment and Work in New Zealand: Proceedings of the Fifth Conference, November 12-13, 1992*, Department of Geography, Victoria University: Wellington.

Curtis, B. (1996), *Producers, Processors and Markets: A Study of the Export Meat Industry in New Zealand*, Unpublished PhD Thesis, University of Canterbury: Christchurch.

Denoon, D. (1983), *Settler Capitalism*, Oxford University Press: New York.

Fligstein, N. (1996), 'Markets as Politics: A Political-Cultural Approach to Market Institutions' *American Sociological Review*, Vol. 61, August, pp. 656-673.

Font, M. (1990), 'Export Agriculture and Development Paths: Independent Farming in Comparative Perspective', *Journal of Historical Sociology*, Vol. 3, No. 4, pp. 329-361.

Friedmann, H. and McMichael, P. (1989), 'Agriculture and the State System: The Rise and Decline of National Agricultures, 1870 to the present', *Sociologia Ruralis*, Vol. 29, No. 2, pp. 93-117.

Gereffi, C., Korzenienic, M. and Korzenienic, R. (1994), 'Introduction', in Gereffi, C. and Korzenienic, M. (eds), *Commodity Chains and Global Capitalism*, Greenwood Press: London.

Harrison, G. (1963), *Borthwicks: A Century in the Meat Trade*, Thomas Borthwicks and Sons: London.

Hartley, M. (1989), *Small Meat Processing Firms Within the New Zealand Meat Industry*, Masters Thesis, University of Auckland: Auckland.

Hayward, D. (1972), *Golden Jubilee: The Story of the First Fifty Years Of the New Zealand Meat Producers' Board*, Universal Printers: Wellington.

Knightley, P. (1981), *The Vestey Affair*, Macdonald: London.

Lawrence, G., Share, P. and Campbell, H. (1992), 'The Restructuring of Agriculture and Rural Society: Evidence from Australia and New Zealand', *Journal of Australian Political Economy*, No. 30, pp. 1-23.

Le Heron, R. (1988), *Reorganisation of the New Zealand Export Meat Freezing Industry: Political Dilemmas and Spatial Impacts*, Massey University: Palmerston North.

McMichael, P. (1984), *Settlers and the Agrarian Question: Foundations of Capitalism in Colonial Australia*, Cambridge University Press: New York.

McMichael, P., and Myhre, D. (1991), 'Global Regulation vs. the Nation-state: Agro-food Systems and the New Politics of Capital', *Capital and Class*, Vol. 43, Spring, pp. 83-105.

Moran, W., Blunden, G. and Bradly, A. (1996a), 'Empowering Family Farms Through Cooperatives and Producer Marketing Boards', *Economic Geography*, Vol. 17, No.1, pp. 22-42.

Moran, W., Blunden, G., Workman, G. and Bradly, A. (1996b), 'Family Farmers, Real Regulation, and the Experience of Food Regimes', *Journal of Rural Studies*, Vol. 12, No. 3, pp. 245-258.

Nolan, J. (1994), *Economic Restructuring and the Viability of the Family Farm*, Unpublished Masters Thesis, Waikato University: Hamilton.

Perry, N. (1992), 'Upside Down or Downside Up?: Sectoral Interests, Structural Change, and Public Policy', in Deeks, J. and Perry, N. (eds), *Controlling Interests: Business, the State and Society in New Zealand*, Auckland University Press: Auckland.

Roche, M. (1992), *Global Food Regimes: New Zealand's Place in the International Frozen Meat Trade: 1883-1935*, Conference of Historical Geographers, University of British Columbia: Vancouver, 15-22 August.

Savage, J. (1990), 'Export Meat Processing', in Savage, J. and Bollard, A. (eds), *Turning It Around: Closure and Revitalisation in New Zealand Industry*, Oxford University Press: Auckland, pp. 59-78.

Smith, P. (1969), *Politics and Beef in Argentina: Patterns of Conflict and Change*, Columbia University Press: New York.

Stinchombe, A. (1961), 'Agricultural Enterprise and Rural Class Relations', *American Journal of Sociology*, Vol. 67, pp. 165-176.

12 Food safety and the New Zealand dairy industry: The politics of Stolle hyperimmune milk

CHRISTINA I. BALDWIN

Introduction

This chapter focuses on a value-added specialised milk powder manufactured in New Zealand for Stolle Milk Biologics International Incorporated (SMBI).[1] This product, known as Stolle hyperimmune milk (SHM), is produced from cows which have been immunised with a vaccine for human pathogens, which induces the formation of specific antibodies in the milk. It is claimed that regular consumption of hyperimmune milk provides health benefits for humans. This chapter questions the validity of the claims, and draws attention to issues such as the non-disclosure of the results of the clinical trials of the product. The key consumer concerns of food safety, traceability and accountability are also discussed.

This discussion is primarily undertaken from the perspective of a New Zealand dairy industry farmer and milk producer, who is critical of the decision to manufacture and market this product. The research is not based on a systematic sociological perspective, but rather an activists' practical and participatory focus on a sphere of activity of the New Zealand Dairy Board, the cooperative which has a shareholding in it (the New Zealand Cooperative Dairy Company), and an associated company, Stolle Milk Biologics International Incorporated, based in Ohio. While attention is mainly focused on one particular product in the New Zealand context, this case study has much wider relevance in terms of food safety, given that (functional) food products are increasingly subject to manipulation by the food industry in order to incorporate additional 'benefits'.

New Zealand dairy industry

The New Zealand dairy industry has an integrated production, processing and marketing structure. It is cooperatively owned and controlled by farmers, with critical links between the significant players in the industry at all key points. Thus, all milk which is produced from dairy cows on New Zealand farms is supplied to twelve cooperative dairy companies in which the suppliers are shareholders. The milk is processed into butter, cheese, milk powder and many value-added products for the consumer, ingredient and food service sectors of the market, by the cooperative companies with access to the latest technology. The single marketing and distribution arm of the industry, owned by the cooperative companies, is the New Zealand Dairy Board (NZDB). The Dairy Board sees the integrated structure of the industry as pivotal to its role of marketing dairy products to customers throughout the world. It makes regular statements of accountability to all New Zealand dairy farmers both through NZDB publications and at dairy industry meetings and conferences. While the single-seller status of the New Zealand Dairy Board is protected by state legislation, it receives no subsidies or financial assistance from the state, and manages its own affairs in all such matters as borrowings, foreign currency management, income and expenditure, and marketing activities.

In its vision statement the Board adopts a highly ethical stance, expressed in a commitment to achieving superior and sustainable returns for its producers, high standards of product quality for its consumers, and environmental integrity for the citizens of countries in which it operates (NZDB, 1996). In practice, information about industry activities, elicited from dairy industry personnel during the research for this chapter, was almost exclusively couched in terms familiar to neoclassical economists or global marketing strategists, and imbued with the tendency to reduce everything to an economic dimension. However, the evidence presented in this case study suggests that it is also necessary to establish strong moral and ethical bonds in order to develop successful economic links between New Zealand and other structures in the global economy, and to maintain the specific relationships with customers and consumers. To put it another way, while high sounding moral principles are frequently alluded to in corporate vision statements, some simple notions to which most citizens subscribe, such as honesty and transparency (particularly in dealing with consumers) do not always seem to be part of the system. This, it can be argued, is the case with Stolle hyperimmune milk.

The origins of Stolle hyperimmune milk

In the 1950s an Ohio businessman named Ralph Stolle took an interest in the previously developed concept of immune milk, and established the Immune Milk Company of America in 1960. Programmes were developed by scientists working at the Stolle laboratories to produce immune milk, to develop a process for spray drying the milk that would not destroy the antibodies, and to establish the composition of the vaccine, and the method of immunisation and quality control procedures for the products. In the late 1980s, the Stolle Research and Development Corporation (SRDC) proposed to develop and license milk biologic products. Approval for the sale of hyperimmune milk for human consumption was granted by the State of Ohio, as long as no medical claims were included on the label. Rather than develop an infrastructure and market milk biologic products in the United States where the scientific and market research had been done, the Corporation looked to New Zealand for the production and processing of the specialty milk. In 1988, a partnership company, Stolle Milk Biologics International, was formed the SRDC and the NZDB. The business of the partnership was referred to as 'health control through passive immunity and immunomodulation' (Beck and Zimmerman, 1989: 98), and the product - the milk powder known as hyperimmune milk - was to be marketed under the brand names of Stolle or Ultralac.

Product development

Stolle hyperimmune milk is obtained from cows which have been immunised selectively by injection against certain bacterial or viral pathogens, or other foreign antigens. Antigen is a term for a factor in the environment capable of stimulating the immune system to produce corresponding antibodies. Antibodies are factors in the immune system which are produced to eliminate, neutralise or 'fight' the antigens. Exposure to an antigen can occur from encountered environmental factors or by vaccination (Beck and Zimmerman,1989: 5). The rationale for immune milk, it is argued by the protagonists, is that 'passive' immune protection is provided from the antibodies or immune substances produced in the cows' milk in response to specific immunisation, to recipients who drink the milk (Beck and Zimmerman, 1989).

In 1989, two members of the SRDC, Doctors Beck and Zimmerman, produced a short paper about Stolle hyperimmune milk (published by the company), which was intended for a diverse audience with varying levels of understanding about immunology and immune milk. They described the immune system as being like an immune 'symphony', claiming that the term symphony denoted 'the harmonious complexity of the immune system'. However, they conceded that they were only beginning to understand the complexity of the human immune system and the immunoregulatory function of milk (Beck and Zimmerman, 1989: 84). It was also admitted that 'work with milk antibodies has only scratched the surface' and that 'we cannot begin to understand the...immune system without a better understanding of the chemical conductors' (Beck and Zimmerman, 1989: 15).

One of the original research experiments involved identifying the spectrum of infectious bacteria which reside in the gastrointestinal tract of rheumatoid arthritis sufferers. The formula for the vaccine which was developed corresponded very closely with this spectrum of bacteria. It included:

> Stapylococcus aureus
> Staphylococcus epidermis
> Streptococcus pyogenes, A. Types 1, 3, 5, 8, 12, 14, 18, 22
> Aerobacter aerogenes
> Escherichia coli
> Salmonella enteritidis
> Pseudomonas aeruginosa
> Klebsiella pneumoniae
> Haemophilus influenzae
> Streptococcus viridans
> Proteus vulgaris, Shigella dysenteriae
> Streptococcus, Group B
> Diplococcus pneumoniae
> Streptococcus mutans
> Cornebacterium, Acne, Types 1 and 2.
> (US Patent Office, 1988; 1993).

This original formula known as Series 100, is produced in the USA and has not changed over the years. It is administered as a sterile vaccine by injecting the cows, which then produce milk with antibodies (Beck and Zimmerman, 1989). It is the resultant milk which is processed by the New Zealand dairy farmers to manufacture hyperimmune milk.

Stolle hyperimmune milk production in New Zealand

Stolle hyperimmune milk is produced on New Zealand farms by farmer/ suppliers of the New Zealand Dairy Company under contract to the New Zealand Dairy Board, which has a 25 percent share in the holding company, Stolle Milk Biologics International. Of the many hundreds of dairy farmers in the Morrinsville district in the North Island, 120 have signed up for the Stolle production program, and this district is a major site for hyperimmune milk production. There are now well over 20,000 cows in the district receiving fortnightly injections of the vaccine. The milk is dried at the Morrinsville site of the New Zealand Cooperative Dairy Company under carefully regulated procedures, so as not to destroy the immune factors in the milk. Strict hygiene and quality procedures are followed from farm to end product. From time to time there is some consumer backlash over the stress placed on cows in the pursuit of higher or differentiated products in the dairy system, but the treatment of the cows in this case conforms to the ethical requirements of the Animal Ethics Committee of the University of Waikato (NZDB, 1996).

Some restrictions in the location of production of Stolle hyperimmune milk exist because of the perishability of fresh milk. However, the expertise and skill requirements are very different for each segment of the process, and a complex web of different participants is now involved along the whole chain from production to consumption. The technology and vaccine are supplied by SMBI, and reconstituted in New Zealand laboratories. The staff of the Livestock Improvement Corporation has an involvement in the animal handling part of the programme. The product is processed by Anchor Products at its Morrinsville factory and packaged by Sachet Packaging in Auckland; both of these are divisions of the New Zealand Dairy Group of Companies. The final product is exported to mainly Asian markets, and marketed by other related companies. Thus, the Taiwan importer is Stolle International (Taiwan) but, New Zealand dairy industry personnel claim only responsibility for the supply, processing and packaging of Stolle hyperimmune milk, and not the marketing.

With such a complex web of participants, any problems which arise with a product such as Stolle hyperimmune milk become more difficult to resolve. For example, while it is usually possible for the industry to trace a hazardous substance, such as a chemical residue in an end product, the problem is more difficult in a production system involving numerous agents at several production and marketing sites, spread across a number of countries. When such complex food safety or marketing questions are raised, the issue is often referred from one location or entity to another, making traceability almost impossible and responsibility difficult to determine.

There are, in addition, other specific issues in the production and marketing of SHM which impinge directly on the New Zealand dairy industry and which have significant implications for its reputation as a producer of environmentally-sound and safe food products. Some of these are discussed in the following section.

New Zealand dairy industry issues

While there are a number of significant issues for the New Zealand dairy industry in relation to Stolle hyperimmune milk, only three will be examined here. The first concerns the validity of some of the claims which have been made for the product; the second relates to the issue of the non-disclosure of the results of clinical trials of the product carried out on human subjects; and the third touches on some issues of importance to the local market.

Natural product or scientific invention?

Dr. Beck, the vice-president and Director of Research of the SRDC, visited New Zealand in May 1997. A meeting was convened at this time by the New Zealand Cooperative Dairy Company, between Dr. Beck, this researcher and other industry personnel. At this meeting, one of the contested issues related to the use of the term 'natural' in the context of the dairy industry and its products. According to Dr. Beck, cows' milk is a natural product for human consumption, and Stolle hyperimmune milk is also a natural product, absolutely safe for human consumption and healthier than a range of other milk products. In contesting this claim, it was suggested to Dr. Beck, that while milk might arguably be a good food for humans, that did not make it 'natural'.

However, even more significant is the fact that elsewhere, Dr. Beck has described hyperimmune milk as an 'invention' (US Patent Office, 1993), while on the labels of Stolle products, it is stated that the cows milk is 'tailored'. Furthermore, Stolle hyperimmune milk is manipulated in many significant ways. For example, the level of antibodies in the milk can be manipulated by altering the vaccine regime (*New Zealand Dairy Exporter*, October 1994). In addition, the composition of the vaccine is deliberately selected by scientific personnel, and is based on substances found in the gut of people with the disease of rheumatoid arthritis. The immunisation process thus occurs as the result of active, not passive, exposure to the antigen. According to Beck, animals are exposed naturally to many harmful antigens, such as carcinogenic, bacterial and viral, in their environment. Normally upon exposure, their immune systems produce antibodies that will neutralize the effect of the antigen. Beck thus argued that whether the animals are exposed to foreign antigens 'naturally' or deliberately by administration, the results are the same as far as that animal is concerned (US Patent Office, 1993). In other words, Stolle hyperimmune milk is 'natural' because cows pick up antigens in their 'natural' environment, therefore injecting them with a more antigens and harvesting these in the milk is also 'natural' and, by association, healthy.

However, it is debatable whether such ideas are normally inherent with the common understanding of 'natural' products. For example, one of the key images that the New Zealand dairy industry promotes overseas is the clean, green environment of New Zealand. Cows are shown in abundant green pastures, grazing beside clear water rippling down from Mount Taranaki's sparkling peak, their udders bulging with fresh milk. The notion of New Zealand's environment as uncontaminated, clean and natural is intrinsic. The idea that cows' milk is naturally good for human health is strongly associated with this image. Beck's notion of cows naturally developing antibodies as a result of living in a contaminated environment and then being injected with a proprietary vaccine for human pathogens to induce the formation of specific antibodies in their milk, is hardly consistent with this picture of a wholesome product derived from a pristine environment.

Of course, if Beck's position does have substance, then milk should not even be pasteurized but sold to the public fresh from the cow. On the other hand, if environmental integrity is valued, why is it necessary to inject cows with additives? In reality, neither position is accurate, and both beg the question of environmental integrity and the nature of the 'naturally' contaminated environment which some cows may inhabit.

Is Stolle hyperimmune milk a food or a drug?

The United States Food and Drug Administration (FDA) considers hyperimmune milk to be a food, unless a medical claim is made on the label. Thus, even though hyperimmune milk may have a spectrum of antibodies different from other milks, because the food value remains the same from one milk product to the next, it is classified as a food so long as the antibodies in the milk are not identified. So, if the manufacturers choose to make a medical claim on the label, the product will be regulated by the FDA. If they choose not to make a medical claim, the food is regulated in the same way as ordinary milk

(Beck and Zimmerman, 1989). There are apparent flaws in this position in that it does not ensure that consumers can identify what the modified content of the food is. It is thus possible for hyperimmune milk to contain an undisclosed range of antibodies, while the consumer has no way of identifying what these antibodies are. Dr. Beck argues that no medical claims are made on Stolle labels because consumers are not yet ready for the medical food concept (Beck and Zimmerman, 1989: 31). As there is a buoyant international market for medical food, other reasons seem likely. If Stolle scientists had been able to establish and demonstrate that hyperimmune milk has proven medical benefits for, or prevents diseases in, humans, it is unlikely that consumers would not be ready to accept this concept.

Non-disclosure of clinical trials

Since the 1960s a number of clinical trials of hyperimmune milk have been undertaken on research animals such as rabbits and mice, and also on humans. Trial work continues in the 1990s. The SRDC, via the New Zealand Dairy Board, provided this researcher with a heavily edited summary of data on these trials. However, the Dairy Board required the researcher to sign a binding non-disclosure agreement before the data was supplied. It is not possible therefore, to provide details of the clinical trials which have been undertaken on Stolle hyperimmune milk. Limited discussion of these trials is available for public examination (see Beck and Zimmerman, 1989), while the results of some other trials are published in nutritional journals (see Golay *et al.*, 1990; Sharpe *et al.*, 1994). Some discussion on the applications of hyperimmune milk within the AIDS community can be found on the Internet.[2] Until trials of Stolle hyperimmune milk are subjected to public scrutiny and scientific debate, any claims about the products must remain unsupported.

What *is* public knowledge, is that most of the trials of Stolle hyperimmune milk were done by Stolle scientists on people with diagnosed diseases or conditions such as psoriasis, rheumatoid arthritis, juvenile arthritis, hyper-cholesterolemia and hypertension. In general, these trials were developed to test immune milk for 'tangible benefits' on people with disease conditions (Beck and Zimmerman, 1989: 32). Some health benefits were reported, but the results were mixed and often inconclusive. Side effects, similar to lactose intolerance symptoms, were not uncommon. One trial was conducted on highly- trained distance runners. Following criticism that these studies were biased, an independent study was conducted in 1984 on 20 healthy men aged between twenty and thirty years. The product was validated for this group

(Beck and Zimmerman, 1989).

It is also clear that there are significant gaps in the research programme. For example, no documentation was located of research having been conducted on normal, healthy children or young adults under twenty years of age, who have consumed SHM regularly over the span of the growing years. It is possible that some young children may be more vulnerable to hyperimmune milk than adults. It is arguable that even the lowest risk is too high for children, especially when it is acknowledged that many features of the immune system remain a mystery. There is no evidence of research having been done on other significant groups, such as pregnant women, specific ethnic groups (some of whom are known to have a higher degree of lactose intolerance than others), or the aged.

Issues in the New Zealand market

While Stolle hyperimmune milk is available for the public to purchase through the dairy industry, it is not yet generally available on the New Zealand market. However, there have been marketing trials which, in many respects, have proved to be as controversial as any other aspect of the production and marketing of the product.

In May 1995, Sachet Packaging, a division of the New Zealand Dairy Group of Companies, in association with the New Zealand Dairy Board, began a trial programme of Stolle hyperimmune milk in New Zealand involving interested volunteers. Upon requesting a sample of Stolle hyperimmune milk through the industry, this researcher was also enrolled on the survey. A questionnaire was provided, requesting information such as the physical reaction of those surveyed following consumption of the product. This trial did not conform to the ethical safeguards which normally regulate research dealing with human participants. For example, no information or guidelines for consumption of the product accompanied the supply. No justification for the project was provided, nor were objectives, indication or information of procedures in relation to informed consent, or the handling of information and materials produced in the course of the research identified (Sachet Packaging, 1995a). Following concerns expressed by this researcher to the Chairman of the New Zealand Cooperative Dairy Company, this trial was halted. However, the trial records were apparently not destroyed, as in December 1995, another survey was mailed out. This survey had been contracted out by Sachet Packaging to a firm of employment consultants. The questionnaire was presented this time in a more sophisticated form, but still had the same flaws as the earlier one. For example, one entire section of this survey was devoted to health and medication, but there was still no minimisation of risk nor safeguards for the participants (Sachet Packaging, 1995b). Such a lack of attention to either the ethical or scientific standards which would be normal requirements of any professional body conducting research involving human participants, was unacceptable.

From the time Stolle hyperimmune milk was produced in New Zealand, interest in the product within the dairy community was high. Anecdotal claims that hyperimmune milk was effective in relation to specific human diseases abounded. But despite the demands of producing hyperimmune milk for the international market, and the awareness of various surveys and studies conducted by the SRDC, the New Zealand dairy industry has not publicly supported health claims about Stolle hyperimmune milk.

Consumer issues

In addition to the issues discussed above, the production and marketing of Stolle hyperimmune milk has raised a number of other issues of general significance to consumers. These are discussed below.

Packaging and labelling

Stolle hyperimmune milk is sold in 20 gram and 45 gram sachets, mainly in Asian markets. It has a shelf life of two years. Each sachet bears the following message:

> Cows are regularly immunised with a proprietary vaccine. Careful processing techniques are employed to ensure the antibodies are preserved.

Nutritional information is also listed. The active antibody content per 45 grams is shown as: Immunoglobulin G 225mg. The outside package also contains instructions for consumption of the product and includes the following message:

> For centuries, humans have consumed cow's milk for its nutritional value. However, for a newborn mammal, milk is more than just a food. Besides vital nutrients, it provides immune factors to help protect against infections in the new environment. Based on this concept, Stolle hyperimmune milk is an innovative new product manufactured in New Zealand where cow's milk is tailored to provide immune-related health benefits to humans. In order to obtain

specific protective factors for humans, cows are immunised with a proprietary vaccine for human pathogens which induces the formation of specific antibodies in the milk. This process is known as 'hyperimmunisation' and routinely produces high titres of specific antibodies in the milk...Regular consumption of Stolle hyperimmune milk provides health benefits and helps promote a healthy gut. Hence Stolle hyperimmune milk drinkers have greater protection from specific pathogens and better regulation of physical functions.

It is not intended in this context to deconstruct the wording on the package, suffice to say that the concerns expressed in this paper about hyperimmune milk are not diminished by the above information. For example, informing the consumer that the product contains Immunoglobulin G does not eliminate concerns about its presence or function. It is also of concern that as long as no medical claims are made on the label, the hyperimmune milk can actually contain any range of antibodies the SRDC selects. As long as the research scientists do not inform the consumer, it is possible to modify the product in any way they choose. The New Zealand Dairy Board is confident that no claims are made about Stolle which do not conform to the marketing regulations of the country of sale, while all claims which are made are scientifically sustainable (NZDB, 1996). However, as it has not been demonstrated that Stolle hyperimmune milk has medical benefits for humans, and it has not been established that the product causes no harm to the target market, non-specific claims protect no-one.

Target market

Although Stolle hyperimmune milk was developed primarily as a medical food for people with specific diseases, it is now marketed as a food suitable for healthy family consumption. However, the product has not been clinically trialed for this target group. The lack of rigorous, long term trials of Stolle hyperimmune milk on groups within the community who are now targeted as consumers, is a matter of great concern. Growing children with developing immune systems are among these groups. There is no clinical evidence that Stolle hyperimmune milk is safe for the very young, very old or pregnant women. The non-suitability of hyperimmune milk as a food for infants is not indicated; in fact, reference to newborns on the package could suggest it is an appropriate infant food. No warning of any contra-indications or interactions with other medications is given on the packaging.

Insufficient knowledge of the immune system

As noted earlier, Dr. Beck has admitted that not much is known about the immune system. That hyperimmune milk can cause an antigen/antibody interaction can only be a matter of speculation. The exact mechanism of action is unknown (US National Library of Medicine, 1996). If partial knowledge and speculation is the real basis for this product, what else is there that remains unknown? As with health problems associated with consumption of other foods, a considerable length of time may need to pass for problems to be indicated, yet any damage to health could be irreversible.

The development of other vaccines

Scientists from the SRDC have selected the range of antigens which determine the composition of the vaccine. Even though quality controls are currently in place, it is possible for the composition of the vaccine to be altered. SRDC scientists concede that different hyperimmune milks have different antibody spectra and therefore have different functional values (Beck and Zimmerman, 1989: 91). They have also introduced research programmes intended to increase knowledge and available technologies in a range of areas. For example, in association with AIDS research, cows have been immunised specifically against cryptosporidia, with specific IgG antibodies against cryptosporidia appearing in the milk in high concentrations. The resultant Stolle milk has been administered to AIDS patients suffering from cryptosporidial enteritis (Beck and Zimmerman, 1989). While this may well be a useful medical food for AIDS patients, the issues for AIDS patients consuming the product are quite specific and different from those for other groups. Not all consumers at this stage are given the assurances that they are not the unwitting target for products not developed for them. As well, Dr. Beck's interpretation of the meaning of 'natural' would allow him unlimited freedom to experiment with any bacteria or viruses he chooses, but present regulations do not require the declaration of all relevant information to consumers. So while new issues relating to human health arise as technology advances, new measures or precautions to regulate these issues are needed to keep up with this advance. Information technology can provide the ability to develop such systems.

Conclusion

Inside the cover of the 1997 Annual Report of the New Zealand Dairy Board is an elegant photograph of New Zealand cows, shrouded in the early morning mists, grazing peacefully. The caption reads:

> New Zealand enjoys an international reputation second to none for the unspoilt nature of its environment. The mission of the Dairy Board is to maximise the sustainable incomes of New Zealand dairy farmers and the key strategy employed to achieve that goal is to add value to products at every stage of the marketing chain. Increasingly the Board is building New Zealand environmental values into the products it markets around the world - qualities which provide a crucial point of difference in the highly competitive international market place.

Environmental integrity is a significant goal, and is difficult to attain even when it is the responsibility of every person within the dairy industry. However, it is interesting that food safety issues were not also cited as desirable goals, along with that of environmental integrity. It is possible that questions about the safety of a food product could provoke a more serious consumer crisis than almost any other issue. The chief executive of the Federated Farmers has recently warned that farmers and producer companies who do not put food safety, traceability and accountability at the top of their agenda are liable to be punished by overseas markets (*Waikato Times*, 2 May 1997). There is no doubt that New Zealand dairy industry personnel do take very seriously their responsibility to ensure that the food they produce does not contain any toxic or hazardous substances. However, if those associated with the New Zealand dairy industry believe that its products are safe and that this is crucial in the international market place, then they should be willing to allow these claims to be exposed to public and consumer scrutiny. Consumer information, including research documentation, should be accessible to farmers, all contributing dairy industry personnel, and to the world-wide customers and consumers of New Zealand dairy products. In the case of Stolle hyperimmune milk, not only is food safety and traceability difficult to establish, but there is also a mis-match between the market for which the products are developed and the markets in which they are being promoted.

As there is clearly huge potential for products in the medical food market, companies such as Stolle Milk Biologics International should support rigorous, independent clinical trials which will stand up to scientific and public scrutiny, thereby firmly establishing the claims made for such products. Stolle Milk Biologics International and others can then be recognised as reliable producers of medical foods, able to sell their products in the medical food market, to consumers who can be confident that they are purchasing safe products for themselves and their families.

Notes

1   I am grateful for the assistance and suggestions of two friends and colleagues, Munikh Clayton, who has specific expertise in social issues for the aged as well as a keen interest in medical foods, food safety and consumer issues, and Sukhjeet Singh-Thandi, also a dairy farmer and shareholder in the New Zealand dairy industry.

2   For example, <http://milksci.unizar.es/aids/aids.html>.

References

Beck, L. and Zimmerman, V. (1989), *Stolle Hyperimmune Milk*, Stolle Milk Biologics International: Ohio.

Golay, A., Ferrara, J-M., Felber, J-P. and Schneider, H. (1990), 'Cholesterol-lowering Effects of Skim Milk from Immunized Cows in Hypercholesterolemic Patients', *American Journal of Clinical Nutrition*, Vol. 52, pp. 1014-9.

New Zealand Dairy Board (1996), *Annual Report*, Wellington.

New Zealand Dairy Board (1997), *Annual Report*, Wellington.

New Zealand Dairy Board, 14 May 1996, personal communication by the Public Affairs Office: Wellington.

New Zealand Dairy Board (n.d.), *New Zealand Dairy Industry Vision*, Wellington.

New Zealand Dairy Exporter (1994), 'Immune Milk Production Expands to Meet Asian Demands', October, pp. 54-55.

Sachet Packaging Limited (1995a), *Stollait Immune Milk Powder Trial Programme*, Auckland.

Sachet Packaging Limited (1995b), *Stollait Immune Milk Powder Trial Programme*, Auckland.

Sharpe, S., Gumble, G. and Sharpe, N. (1994), 'Cholesterol-lowering and Blood Pressure Effects of Immune Milk', *American Journal of Clinical Nutrition*, Vol. 59, pp. 929-34.

US National Library of Medicine (1996), *Cryptosporidium Immune Whey Protein Concentrate (Bovine)*, Bethesda, Maryland.

US Patent Office (1988), *4732757 Prevention and Treatment of Rheumatoid Arthritis*, Washington D.C. <http://patent.womplex.ibm.com/details?patent_number =4732757>.

US Patent Office (1993), *5194255 Antihypertensive Hyperimmune Milk, Production, Composition, and Use*, Washington D.C. <http://patent.womplex.ibm.com/details?patent _number=5194255>.

Waikato Times (1997), 'Food Safety Issue a Risk: Farm Leader', 2 May: Hamilton.

13 Fertiliser and sustainable land management in pastoral farming, Northland

GREG BLUNDEN AND BEN BRADSHAW

Introduction

Agriculture has been subject to regulatory and financial support by governments in many countries, especially since the early 1960s when major trading blocs began to emerge and food production surged in western countries.[1] Indeed, state intervention in the agricultural sector of most national economies has traditionally been, and often still is, greater than that of any other sector (Robinson, 1989). Support has taken many forms, including public scientific research, farm finance and development schemes, production price support mechanisms like the Common Agricultural Policy (CAP) in the EU and supplementary minimum prices (SMPs) in New Zealand, and subsidies for the purchase, transport and/or spreading of fertiliser and for other farm inputs. In essence, by providing this support, government intervention produced the risk-free environment necessary to encourage massive investment (Bowler, 1985), which thereby enabled vast expansions in production capability. Hence, as in other countries, output increased greatly in New Zealand during the decades after the Second World War.

While many of these support mechanisms remain in most western countries, almost all forms of support to land-based production in New Zealand, including all indirect and direct price supports for fertiliser, were removed from 1984 onwards. New Zealand agriculture now boasts the lowest level of government support of any country in the OECD. As measured by the producer subsidy equivalent (PSE), a commonly used indicator of support which expresses in percentage terms the total value of state transfers and support to agriculture relative to the total value of agricultural output, support to the agricultural sector in New Zealand was just three percent in 1996, compared with an average 43 percent in the EU (OECD, 1997) (see Figure 13.1).

In analysing the issue of fertiliser and sustainable land management in pastoral farming, it is important to note that New Zealand pastures are based on introduced species, principally rye grasses and clovers, which have a high demand for nutrients (Williams and Haynes, 1991). Once developed, pastures require maintenance applications of nutrients and trace elements, particularly phosphorus, sulphur, potassium, molybdenum, copper and boron to maintain fertility (During, 1984). There is a direct relationship between fertiliser use and pasture and animal production (O'Connor *et al.*, 1990), and applications of key elements are essential components for clover-based pasture systems to maintain nitrogen fixation rates and the growth of productive pasture (King and Krausse, 1995).

For pastoral farmers in New Zealand decisions about fertiliser use have, since 1984, been based on economic as much as biological grounds. To a certain extent, this was always the case as a direct, long-run correlation exists between the amount of fertiliser used and the returns to the various farming systems. However, economic criteria have become more important with the continued decline in nominal and real pastoral incomes since the mid-1980s, and this has tended to cause farmers to reduce fertiliser use. New Zealand farmers are aware of the direct relationship between lower fertiliser application

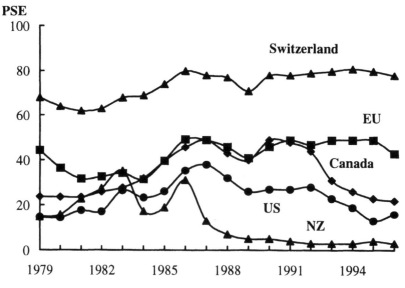

Figure 13.1 Producer subsidy equivalent (PSE), percentage
Source: OECD (1997).

rates and lower pasture production levels, but the economics of pastoral production in post-1984 New Zealand have tended to preclude alternative strategies. It is this issue, and the relationships that lie behind it, that we focus upon here - how and why pastoral farmers have altered their fertiliser practices since the effective cut-off of government support, and what this means for sustainable land management.

This analysis is undertaken at several scales. First, fertiliser use at the national scale is reviewed using aggregate data from Statistics New Zealand. This is complemented by a review of changing fertiliser expenditure at a regional scale, using data from a *typical* Northland sheep and beef farm. We intersperse these with a commentary on the changes to the regulatory system, market prices and farm income to demonstrate their interdependencies with fertiliser use. Second, we draw on three data sets to present farmers' views on these changes. One is a postal survey to the members of Northland Federated Farmers (Blunden and Cocklin, 1995; Blunden *et al.*, 1996). Another is ethnographic information from 50 farms in central Northland. The farms range from horticultural units to small scale beef and sheep farms and small dairy farms to large-scale farms of both pastoral types, which include in some cases, deer and, in many cases, exotic forestry (Scott *et al.*, 1997). The third data set is a selection of Northland beef and sheep farms whose owner-operators were interviewed in 1996 (Bradshaw, forthcoming).

Regulation, fertiliser use and pastoral farming in New Zealand

New Zealand's modern history is related inextricably to the development of land for pastoral farming. Legislation has been an essential tool in the development of land-based production and, indeed, farmers dominated many New Zealand governments until recent times. New Zealand's family-based agriculture was nourished by a diet of centralised marketing, a huge public science effort, and various other forms of direct and indirect support from central government. Farming was seen as the backbone of the country, and New Zealand's prosperity relied on its (subservient) role of food producer to the United Kingdom, such was the country's reliance on Britain for export income. All this occurred in a unicameral state where egalitarianism reigned supreme and government managed the economy to an extent rarely seen elsewhere in the western world. For example, access to, and prices for, public services and essential goods such as petrol were equalised wherever possible, despite a physical geography that made the cost of providing such goods and services hugely variable. New Zealand was the epitome of the welfare state.

When Britain joined the EEC and the global economy entered a period of recession in the 1970s, New Zealand's response was to attempt to continue as before. The New Zealand government borrowed abroad to maintain the standard of living, and turned to farming and *Think Big* investment projects to make the country more self-reliant and to lessen foreign exchange requirements. Two major schemes were devised during the 1970s to expand agricultural production and so help New Zealand to trade its way out of economic difficulties. Land development grants and concessions were made available during the mid-1970s and supplementary minimum prices (SMPs) were introduced in the late 1970s. Both schemes were aimed at expanding production from hill country farms by bringing relatively unused and marginal land into production. These schemes were complemented by a fertiliser subsidy program that was initiated in 1965 and used aggressively in the late 1970s to make possible below-cost purchase, transport and spreading of fertiliser and other topdressing such as lime. For example, in 1983, when total support to pastoral agriculture equalled NZ$913 million, land development grants in tandem with interest and tax concessions totalled NZ$244 million, SMPs totalled NZ$438 million and the various fertiliser subsidies amounted to NZ$44 million (NZMAF, 1996). The combined effect of these various programs was to increase fertiliser consumption dramatically between 1975 and 1983 (see Figure 13.2).

1984 was a watershed year in New Zealand's social and economic history. The election of the Fourth Labour Government marked the beginning of the dismantling of the welfare state through such measures as the corporatisation and privatisation of government functions, and the introduction of the user-pays principle for services that were formerly *free*. Keynesianism was replaced by the monetarism of Thatcher and Reagan and the neoliberalism of the New Right with its emphasis on individualism, market forces and 'the level playing field'. A key element of the policy changes was the cutting of most, if not all, subsidies to the agricultural sector. Farmers were an obvious target for the newly-elected Labour Government as they were outside the Labour Party's natural constituency. Further, they were an easily-identifiable special interest group because they received direct support to encourage production, and thereby were guilty of the 'crime' of distorting markets. Besides this, the previous National government had already programmed the phasing out of SMPs but had not been able to implement the policy change because of the early calling of an election; hence, Labour was simply following through on National's plans.

In the pre-1984 period, many subsidies, such as those for fertiliser use and transport, were available to all farmers generally. However, the bulk of land development grants and SMPs went to hill country farms that produced wool, lambs and steers for fattening. Indeed, the purpose of the subsidy was to develop this marginal hill country and expand production past the then-existing limits. Consequently, these farmers felt the effects of the policy changes most dramatically after 1984 as the SMPs were discontinued, along with land development grants and concessions, fertiliser subsidies and other less direct forms of government support. Moreover, the effects of these changes were magnified by policy changes in the broader economy, including the floating of the New Zealand dollar and the removal of controls on domestic interest rates early in 1985. As a result, beef and sheep farmers cut fertiliser use dramatically in the face of sharply decreased income and land values and much higher input costs (primarily interest charges). This accounts for the precipitous dip in Figure 13.2 in 1985. The application of fertiliser, which typically accounted for one-third of seasonal expenditure on farms through the 1970s, was curtailed drastically as a result of the removal of subsidies, adverse terms of trade, and regulatory change.

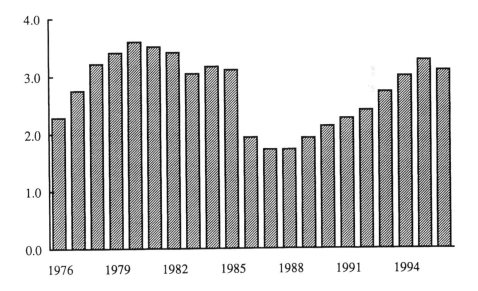

Figure 13.2 Fertiliser use in New Zealand, millions of tonnes
Source: Statistics New Zealand (various years).

Dairying did not experience such a devastating combination of negative factors, partly because SMPs were available only to sheep and beef producers, and partly because the dairy industry had already phased out certain government support mechanisms. Indeed, the contrasting fortunes of beef and sheep producers and dairy farmers have influenced the fertiliser statistics ever since; dairy farmers were primarily responsible for the steady increase in fertiliser use after 1988 towards the levels that prevailed prior to the removal of SMPs in 1985.

Fertiliser use as an economic decision

Pastoral farmers' fortunes declined during the recent period of restructuring in New Zealand, particularly for those raising sheep and beef (Figure 13.3). The decline in nominal and real milk-fat and wool prices is a trend that is mirrored for other long-standing export produce such as meat and apples, as well as for newer produce lines such as kiwifruit. It was declining prices and poor terms of agricultural trade that led to many of the subsidies that were introduced progressively from the early 1960s. Indeed, subsidies, and particularly SMPs in the early 1980s, may have represented a partial reprieve from the cost-price squeeze that sheep and beef farmers had to deal with from the early 1960s until the deregulation of the New Zealand economy which began in 1984. Apart from the minor subsidies put in place in the 1960s and early 1970s, farmers were largely left out of the regularly-renewed corporatist contract between the state and labour that nurtured the welfare state and New Zealand's import substitution industries in the period after the Second World War. Farmers had to deal with the higher import costs that resulted from the corporatist contract, costs which made their operations increasingly less profitable.

Figure 13.3 not only shows the overall decline in the real prices for the outputs of pastoral agriculture, but also reveals the significant variability in prices that is typical of normal price cycles for these commodities. In fact, prices in 1996 reached their lowest ever levels, and this has once again caused farmers to cut back on fertiliser and other unnecessary expenditures. As one Northland sheep and beef producer said,

> We are back in that scenario now [the 1986 crisis]...you have to cut back right across the board. You have to budget, we budget fairly strictly, and you budget for those expenditures which you can't do anything about, like debt servicing,

rates, insurances...personal expenditure, animal health. So there is fixed costs which you can't avoid; so they're the first bunch you add in. Then whatever's left is divided into priorities...and maintenance tends to be low on the scale of importance.

Indeed, all the beef and sheep farmers interviewed reported a significant decrease in fertiliser use and property development (e.g. new fencing, pasture development) in the immediate post-1984 period, as well as during the more recent downturn. In addition, most beef and sheep farmers reported that general maintenance of existing fencing was discontinued if it required money.

The differential fluctuations in incomes for pastoral farmers in New Zealand can be appreciated by considering the indices in Figure 13.4. Average sheep and beef farm incomes declined from 832 in 1985 to just 329 by 1986 and stood at 473 in 1994 (NZMWBES, 1996). In the dairy sector, the index declined only slightly from 969 in 1985 to 893 in 1994, although this apparent consistency hides considerable variation in income over the decade (Livestock Improvement Corporation (LIC), 1996).

The variability in fertiliser use resulting from uneven and generally

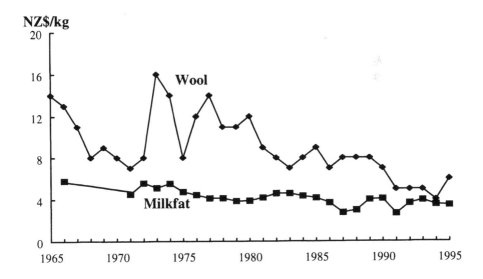

Figure 13.3 Real wool and milk fat prices
Source: Statistics New Zealand (various years); LIC (various years); NZMWBES (various years).

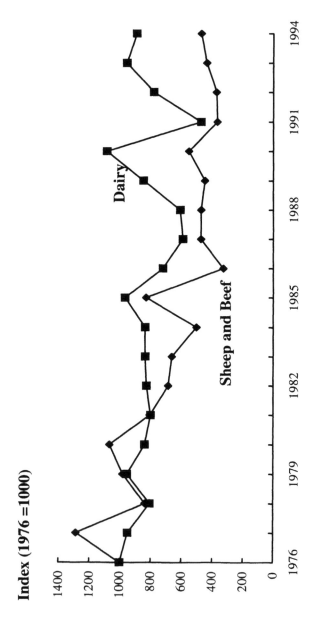

Figure 13.4 Pastoral income indices
Source: NZMWBES (1996); LIC (1996).

depressed returns is reflected in the expenditure data derived from MAF's regional farm modelling of Northland sheep and beef producers. A model or typical farm (Figure 13.5) is developed on an annual basis for each region and sector of New Zealand agriculture, based upon consultation with MAF officials and regional agribusiness, and a sampling of 8-20 real farms. The Northland sheep and beef farm model for the years 1983 to 1996 reveals the highly variable and generally insufficient expenditure on fertilisers and other inputs during the period of restructuring (Figure 13.5). The background data in Figure 13.5 are the (national) average amount spent on fertiliser, lime and seeds by sheep and beef farmers, expressed in 1995 dollars. Both sets of figures closely reflect the pattern of national fertiliser use depicted in figure 13.2; a sharp decline after 1984, and a slight recovery since, but still below 1983/84 levels.

The amount of fertiliser applied by modern conventional farmers always reflects economic factors as well as biological necessity. The balance tilts towards economic factors when farmers suffer a decline in their profitability,

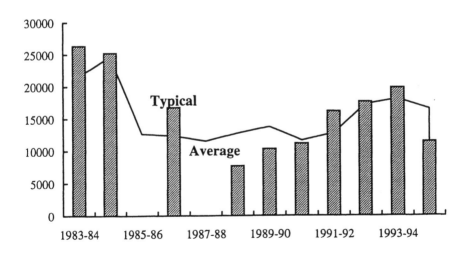

Figure 13.5 'National average' and 'Northland typical' expenditure on fertiliser by sheep and beef farmers, 1995 NZ dollars

Source: NZMWBES (various years); NZMAF (various years).

whether this is due to cost-price squeezes, the macro-economic environment, adverse changes in government policy, or long term declines in real per-unit returns to farming. Conversely, more fertiliser is applied when profits are acceptable or increasing, and increased use of fertiliser occurs at an aggregate level when farmers expand production in response to market signals, or changes in government policy that have the same effect. Since 1984, it has been a combination of negative factors, not just the loss of subsidies, which has made the adjustment process so severe for hill country farmers. The coincidence of these factors made the use of fertiliser prohibitively expensive for many pastoralists.

While these data effectively convey the quantitative change in fertiliser use in post-1984 New Zealand, they do not express some of the qualitative changes that have occurred over the past decade. In addition to overall reduced use, the increased relative cost of fertilisers has prompted farmers both to alter their spreading methods and to employ less expensive varieties. Fertiliser use is now generally more reflective of soil and pasture requirements. Rather than spread equal (and often excessive) quantities of fertiliser across all sections of a property regardless of need, many farmers now determine the specific needs of specific paddocks or sections, and apply fertiliser as needed. As one Northland beef farmer explained:

> Of course now, we're a little bit more sensitive about the cost...and so we're more specific. That is why we soil test and we blood test the cattle, and look carefully at what is needed.

While this management change has been prompted by the increased relative costs of fertiliser *vis-a-vis* farm income, it has been facilitated by new technology that provides cost-effective testing of soil and pasture conditions. The combination of economics and technology has also been central to the decision by many farmers to use alternative forms of fertiliser. For example, many sheep and beef farmers changed to reactive rock phosphate (RPR) from superphosphate (or 'super') during the late 1980s. For many, this change was made because of RPR's reduced cost and its effectiveness as a slow release fertiliser. However, for some, the decision to switch fertilisers also reflected concern over health and the environment. Such was true for this beef and sheep farming couple:

> M: We haven't used any...super for years and years and years...
> F: Going back 13 years isn't it?

M: We were the first farm, probably in New Zealand...that flew RPR on, and ...it just caused a bloody sensation...because of economics [and] animal health...and I realised [what] it was...well, I didn't like super, well I didn't know what super did, but I didn't like it when I handled super. Like I just think that when you go and shovel it...it just seems an obnoxious kind of substance to be putting on something...because...it would stink.

As with the decision to both reduce fertiliser use and to apply it in a more precise manner, the decision by many farmers to use alternative forms of fertiliser, including some organic varieties, has produced environmental gains which were not necessarily intended. These gains, as well as some of the other effects of altered fertiliser use in post-1984 New Zealand, are reviewed in the next section.

The implications of reduced fertiliser use

Over the past decade especially, a greater awareness has formed about the impact of land-based production on the environment and questions have been asked about the sustainability of existing land management practices. For example, Williams and Haynes (1991) argue that pastoral land management in New Zealand is not sustainable in the long term because of the energy and fertiliser imbalance in the pasture system. Underpinning this conclusion is their recognition of the direct relationship between fertiliser use and pasture and animal production (O'Connor *et al.*, 1990). Williams and Haynes concluded that the productivity levels achieved in hill country farming during the subsidised periods of the 1970s and 1980s were unsustainable without the subsidies that existed. Increased erosion (probably because of the development of marginal land) emphasised this unsustainability.

Production subsidies have long been implicated as a contributing cause of environmental degradation in commercial agriculture (e.g. Abler and Shortle, 1992; Body, 1984; Bowler, 1985; De Wit, 1988; OECD, 1989, 1993; WCED, 1987). In particular, input subsidies, which reduce the cost of chemicals, irrigation or even credit, have been targeted for encouraging input intensification and the conversion of marginal land. For example, in separate analyses of the relationship between levels of state assistance and production intensity for a number of national agricultural sectors, Anderson and Strutt (1996) and Lewandrowski *et al.* (1997) observed increased use of environmentally-deleterious inputs with increases in government assistance as measured by the PSE. It is upon this basic correlation that the OECD (1993), among others, contends that environmental benefits can be achieved through reduced state support of agriculture.

The evidence from post-1984 New Zealand, especially in the short term, appears to support this contention. For example, Reynolds *et al.* (1993) demonstrate that the removal of subsidies has produced a number of beneficial environmental spin-offs, not the least of which has been the water quality improvements resulting from reduced (and more precise) fertiliser use. This claim is undoubtedly true; indiscriminate and excessive fertiliser use in the past often created detrimental environmental effects. In tandem with the exorbitant fiscal costs of fertiliser and other subsidies, many New Zealand farmers themselves have come to see their previous use of fertiliser as inappropriate. As two Northland sheep and beef farmers explained:

> One of the bad things about subsidies is because farming costs are tax deductible, say you've had a pretty good year...at the end of the year your accountant says you're going to be paying 60 cents in the dollar tax in those days....[so you think] I'd better spend it on something, so [one] thinks about fertiliser, throws it all on, and it didn't need it anyway. The farm doesn't benefit and the taxpayer doesn't benefit.

> We put it on because we could pay for it, and it was almost like a populist thing, you've got to put fertiliser on; the more you put on the better, regardless.

With the loss of the fertiliser subsidy and, perhaps more importantly, the significant decline in farm income in the post-1984 period, farmers have come to test for the nutrient needs of soils, pastures and livestock so as to use fertiliser and other inputs more selectively. However, it is clear that the impetus for these changes has been economic rather than environmental (see Blunden and Cocklin (1995) and Blunden *et al.* (1996) for further discussion of these priorities), and this raises other concerns over the sustainability of production systems in post-1984 New Zealand.

While reduced fertiliser use may have produced gains for the natural environment, these have come at the cost of pasture production. Indeed, the relationship between fertiliser use and production levels was brought clearly into focus during the recent period of restructuring in New Zealand when most farmers' incomes declined. The effects of the decision taken by many pastoralists to decrease fertiliser use in the post-1984 period quickly became apparent to those within the industry. As one hay contractor noted:

The first thing that they do is they knock off the fertiliser and you've only got to drive round the valley and you can see the farms that have knocked off the fertiliser.

The effects of cutting-back on fertiliser varied considerably with the soil type on the farm; deficiencies showed-up within one season on some sandy soils in Northland, while other farmers in the region were able to maintain reasonable production for some years on much reduced applications of fertiliser. Of course, the future productivity of the soil varies with the degree to which fertiliser is reduced. Many farmers throughout New Zealand cut back to what they called 'maintenance' levels immediately after subsidies ceased in 1985. For one Northland beef and sheep farmer, the combination of the downturn in sheep prices and high interest payments caused him, by 1987, to cut back on discretionary expenses because of income loss;

> Fertilizer was barely at maintenance [levels], sometimes 18 months between dressings.

As established earlier, New Zealand pastures are heavily dependent on nutrients and certain trace elements in order to maintain fertility. Williams and Haynes (1991) cite O'Connor *et al.* (1990) in suggesting that the effects of negative nutrient balance, based on trials on developed pasture on a typical North Island hill country sheep farm, are cumulative: limited change in pasture production was noticed in the first year after fertiliser application ceased; however, production declined by 24 percent after three years and 50 percent after six years; in turn, stocking rates declined, animal health problems intensified, and soil phosphorus levels declined by over 30 percent.

Most farmers who significantly reduced their fertiliser use understand the direct relationship between fertiliser use, grass growth and stock performance. It is not that they won't fertilise, it is that they cannot afford to fertilise;

> It's basically due to lack of fertiliser. If you don't put fertiliser on, you don't get grass growth; so you blame the weather...if your farm is in good heart, then [during periods of] adverse weather, it [will do] better, than if it's not [in good heart];

> Fertiliser is critical - no money can be made without it; and off-farm income and investments are critical in this current period of downturn;

It's impossible to fatten cattle without fertiliser, you've got to have them on fertilised country otherwise you're in trouble. You just can't do it, and the farm will go back - they go back anyway if you don't. And this is the trouble with the North, there are a lot of farmers like - I keep saying if we didn't have our other interests, our other income we would be struggling;

[When you stop applying fertiliser], that's the beginning of the end.

The reduction in fertiliser use has come at a cost in terms of farm productivity, and this may have implications for the sustainability of farming systems in post-1984 New Zealand. Indeed, the evidence that we have assembled from Northland suggests that many sheep and beef operations have become caught in a vicious cycle of declining fortunes, whereby the loss of subsidies and reduced income has prompted a reduction in the use of fertilisers and other inputs, leading to less pasture growth, less weight gain on livestock, reduced stocking rates, reduced output, and hence further reduced income. In this regard, these operations tend to fit a pattern which is common to cash-crop production systems within developing economies. These systems, and in particular their environmental impacts, have been a focus of considerable research under the banner of political ecology. The principle finding of this body of research, and the one that formed the central argument of the influential 1987 Brundtland report (WCED, 1987), is that exploitative or destructive resource use is a product of shortened planning horizons caused by impoverishment (e.g. Ashby, 1985; Blaikie, 1985; Blaikie and Brookfield, 1987; Collins, 1986; Hwang *et al.*, 1994; Painter and Durham, 1995). With increasing poverty, farmers tend to sacrifice long term objectives and make more intensive use of available resources, or as Johnson and Lewis (1995: 291) suggest, they 'adjust upward the upper limit to which a system can be exploited'. When the system's resilience limit is reached, productive output can become erratic. Lower yields imply growing poverty which, in turn, can result in further mining of the land; a vicious cycle is born (De Janvry and Garramon, 1977).

The environmental implications of a similar downward cycle have become apparent in some New Zealand pastoral farming systems (see Smith and Saunders, 1995; 1996). With fertiliser use below maintenance levels, ground cover density has been greatly reduced, thereby allowing for increased soil exposure and erosion. In general, farmers who are unable to secure sustainable returns from production, naturally give less attention to the environmental well-being of their properties; or as Johnson and Lewis (1995:

295) suggest, 'when people and place collide in a struggle for survival, it is the norm that the quality of the resource base is the first victim'. In this regard, reduced fertiliser use in post-1984 New Zealand may account for an actual environmental gain in terms of water quality improvements, but it may also indicate that the system is no longer able to sustain itself.

Conclusion

It is clear from the evidence presented in this paper that government regulations and support measures for agriculture, and in particular input subsidies like those for fertiliser, have the capacity to alter resource use among individual farmers. The production-oriented regulatory regime that was established in pre-1984 New Zealand encouraged farmers to use fertiliser and other inputs in an excessive and indiscriminate manner. By removing these input subsidies, as well as a number of other subsidy programs, farm-level fertiliser use, especially amongst sheep and beef farmers, declined significantly. This has been read by many (e.g. Reynolds *et al.*, 1993) as an indication of improved environmental well-being in post-1984 New Zealand. In itself, this claim is accurate. However, some of the evidence presented in this paper suggests that reduced fertiliser use may also be indicative of overall economic, social and environmental decline.

The economics of sheep and beef farming since the mid-1980s are such that this particular farming system appears to be no longer viable at the present scale. Based on the Northland cases, it is clear that the land is now in poorer condition, farm families are under high levels of stress due to low incomes, and there is no capital available to apply maintenance levels of topdressing or expand farm size in order to cope with falling per unit returns. Many sheep and beef farms have regressed physically, their operators have become lifestyle farmers (Blunden, 1997) dependent on off-farm income to maintain ownership, and many farmers have simply sold the farm. In many of these cases during the 1990s, plantation forestry has replaced pastoral farming. This latter scenario has been a common observance for the Mangakahia hay contractor quoted earlier:

> [The land] can stand it for a couple of years, but sooner or later if they don't fertilise it, then in come the forestry boys and, see, that's why that Houto country has been sold, those farms up the Houto.

Not only does an inability to fertilise have implications for the state of the physical environment and reflect the economic situation facing farmers, but it is also an indicator of more systemic change, in this case, a shift from a pastoral farming system to a forestry system. The sustainability of the new land use is another question entirely.

The ability of farmers to stay on the land, to sustain themselves in the short, medium and long term, is inherently related to the prices they receive for their outputs. Farmers are able to withstand short-term fluctuations in their income, as in any business. Yet, no-one can sustain falling incomes indefinitely. The process that begins this is land use competition and the outcome is often land use change. In the region under investigation, the result was a decline in the number of beef and sheep farms, offset by an increase in the number and size of dairy farms, an increase in the area in plantation forestry, and a change in the rural population (with farmers leaving localities and being replaced by lifestylers and forestry workers in some cases). The physical sustainability of the land is a side issue compared to issues of land use competition.

Herein lies a key to understanding the sustainability question; sustainability, at least at the individual farm level, is as much about economic and social well-being as it is about environmental integrity. The overall message is that sustainable land management is impossible to achieve if the managers of the land are impoverished. This finding echoes that of Smith and Saunders (1995, 1996), as well as that of much research in political ecology. Without economic stability, conservation strategies cannot be implemented to their best advantage. As poverty increases, farmers may make more intensive use of their land, let the land go (in terms of maintaining its condition) or mine the soil. Hence, decreased fertiliser use does not, in itself, lead to increased environmental sustainability, especially if it also indicates a farmer's inability to properly maintain the land.

Note

1   This paper was written under the auspices of the project *Definition and Analysis of Sustainable Land-based Production*, which is funded by the Foundation of Research, Science and Technology (Contract No. UOA 509).

References

Abler, D. and Shortle, J. (1992), 'Environmental and Farm Commodity Policy Linkages in the U.S. and the EC', *European Review of Agricultural Economics*, Vol. 19, No. 2, pp. 197-217.

Anderson, K. and Strutt, A. (1996), 'On Measuring the Environmental Impact of Agricultural Trade Liberalisation', in Bredahl, M., Ballenger, N., Dunmore, J. and Roe, T. (eds), *Agriculture, Trade and the Environment: Discovering and Measuring the Critical Linkages*, Westview Press: Boulder, pp. 151-172.

Ashby, J. (1985), 'The Social Ecology of Soil Erosion in a Columbia Farming System', *Rural Sociology*, Vol. 50, No. 3, pp. 377-396.

Blaikie, P. (1985), *The Political Economy of Soil Erosion in Developing Countries*, Longman: London.

Blaikie, P. and Brookfield, H. (1987), *Land Degradation and Society*, Methuen: London.

Blunden, G. (1997), 'Assessing the Sustainability of Land-based Production: Pastoral Farmers in Northland, New Zealand', in Singh, A. (ed.), *Land Resource Management*, B. R. Publishing: Delhi.

Blunden, G. and Cocklin, C. (1995), *Farmers' Perspectives on Sustainable Land Use and Rural Communities in Northland*, Working Paper No. 1, Department of Geography, University of Auckland: Auckland.

Blunden, G., Cocklin, C., Smith, W. and Moran, W. (1996), 'Sustainability: A View from the Paddock', *New Zealand Geographer*, Vol. 52, No. 3, pp. 24-34.

Body, R. (1984), *Farming in the Clouds*, Temple Smith: London.

Bowler, I. (1985), *Agriculture Under the CAP: A Geography*, Manchester University Press: Manchester.

Bradshaw, B. (forthcoming), *Resource Use Response to Subsidy Removal in Commercial Agriculture*, Unpublished PhD Thesis, University of Guelph: Guelph.

Collins, J. (1986), 'Small Holder Settlement of Tropical South America: The Social Causes of Ecological Destruction', *Human Organization*, Vol. 45, No. 1, pp. 1-10.

De Janvry, A. and Garramon, C. (1977), 'The Dynamics of Rural Poverty in Latin America', *Journal of Peasant Studies*, Vol. 4, pp. 206-216.

De Wit, C. (1988), 'Environmental Impact of the CAP', *European Review of Agricultural Economics*, Vol. 15, No. 2/3, pp. 283-296.

During, C. (1984), *Fertilisers and Soils in New Zealand Farming* (third edition), Government Printer: Wellington.

Goh, K. and Nguyen, M. (1991), 'Fertiliser Needs for Sustainable Agriculture in New Zealand', in Henriques, P. (ed.), *Proceedings of the International Conference on Sustainable Land Management*, Napier, New Zealand, November 17-23.

Hwang, S.W., Alwang, J. and Norton, G.W. (1994), 'Soil Conservation Practices and Farm Income in the Dominican Republic', *Agricultural Systems*, Vol. 46, pp. 59-77.

Johnson, D. and Lewis, L. (1995), *Land Degradation: Creation and Destruction*, Blackwell: Oxford.

King, J. and Krausse, M. (1995), *The Impacts of Land Use Change in Wairoa*, Landcare Research Institute New Zealand: Wellington.

Lewandrowski, J., Tobey, J. and Cook, Z. (1997), 'The Interface Between Agricultural Assistance and the Environment: Chemical Fertiliser Consumption and Area expansion', *Land Economics*, Vol. 73, No. 3, pp. 404-427.

Livestock Improvement Corporation (LIC) (various years), *Dairy Statistics*: Hamilton.

New Zealand Ministry of Agriculture (NZMAF) (1996), *Situation and Outlook for New Zealand Agriculture*: Wellington.

New Zealand Ministry of Agriculture (MAF) (various years), *Farm Monitoring Report - North Region*: Hamilton.

New Zealand Meat and Wool Boards Economic Service (NZMWBES) (various years), *Annual Survey of the Sheep and Beef Industry*: Wellington.

New Zealand Meat and Wool Boards Economic Service (NZMWBES) (1996*), New Zealand Sheep and Beef industry 1995-96*, Paper No. G2121: Wellington.

O'Connor, M., Smart, C. and Ledgard, S. (1990), 'Long Term Effects of Withholding Phosphate Application on North Island Hill Country: Te Kuiti', *Proceedings of the New Zealand Grasslands Association*, Vol. 51, pp. 21-24.

Organisation for Economic Cooperation and Development (OECD) (1989), *Agricultural and Environmental Policies: Opportunities for Integration*: Paris.

Organisation for Economic Cooperation and Development (OECD) (1993), *Agricultural and Environmental Policy Integration: Recent Progress and New Directions*: Paris.

Organisation for Economic Cooperation and Development (OECD) (1997), *Agricultural Policies, Markets and Trade: Monitoring and Outlook*: Paris.

Painter, M. and Durham, W. (1995), *The Social Causes of Environmental Destruction in Latin America*, University of Michigan Press: Ann Arbor.

Reynolds, R., Moore, W., Arthur-Worsop, M. and Storey, M. (1993), *Impacts on the Environment of Reduced Agricultural Subsidies: A Case Study of New Zealand*, MAF Policy Technical Paper 93/12, New Zealand Ministry of Agriculture and Fisheries: Wellington.

Robinson, K.L. (1989), *Farm and Food Policies and their Consequences*, Prentice Hall: Englewood Cliffs.

Scott, K., Park, J., Cocklin, C. and Blunden, G. (1997), 'A Sense of Community: An Ethnography of Rural Sustainability in the Mangakahia Valley, Northland', Occasional Paper No. 33.

Smith, W. and Saunders, L. (1995), 'Agricultural Policy Reforms and Sustainable Land Management: A New Zealand Case Study', *Australian Geographer*, Vol. 26, No. 2, pp. 112-118.

Smith, W. and Saunders, L. (1996), 'Agricultural Sustainability: The Cost of the 1984 Agricultural Policy Reforms', *New Zealand Geographer*, Vol. 52, No. 1, pp. 21-28.

Statistics New Zealand (various years), *Agricultural Statistics*: Wellington.

Williams, P. and Haynes R. (1991), 'New Zealand's Pastoral Farming; How Biologically Sustainable is it?', *Proceedings of the International Conference on Sustainable Land Management*, Napier, November 17-23 pp. 284-291.

World Commission on Environment and Development (WCED) (1987), *Our Common Future*, Oxford University Press: Oxford.

14 Doing good, doing harm? Public debate about Rabbit Calicivirus Disease in New Zealand

GERARD FITZGERALD AND ROGER WILKINSON

Introduction

European rabbits (*Oryctolagus cuniculus*) are New Zealand's second major vertebrate pest problem after brushtail possums (*Trichosurus vulpecula*). The history of rabbits in New Zealand has been summarised in Gibb and Williams (1994). In this chapter, we describe the social history of rabbit calicivirus disease (RCD) in New Zealand, an introduced disease that promised to revolutionise rabbit control. The story is a mixture of scientific 'truths', media images, public views, political realities, and farmers' actions. Sometimes it is hard to tell whether good or harm is being done, and to identify who is doing what to whom. In an attempt to clarify this, we begin by describing the background events and provide a chronology of the debate on the introduction of RCD. Following this, we discuss briefly the results of our focus groups and research surveys, which indicate what the public actually thought about the issue. Then we describe the media coverage of RCD, and conclude with some observations and reflections.

Chronology of the debate

The recent history of rabbit control in New Zealand begins in the 1980s, with the restructuring of the economy through the introduction of a user-pays market system. In 1989, the various local pest destruction boards around the country were wound up, and responsibility for rabbit control passed to newly-created regional councils (and in some areas, other local bodies), which were made responsible for resource management (including pest management). The

Biosecurity Act 1993, and the Regional Pest Management Strategies produced under that Act, introduced a user-pays regime for rabbit control. Under this scheme, landholders were charged the true costs of such control, at a time when other farm subsidies were being removed, farming costs were rising and returns falling. The greatest impact was felt in the rabbit prone areas of Canterbury and Otago.

Myxomatosis was seen as a potential solution to the rabbit problem, and the introduction of the disease and a flea vector to New Zealand was considered several times during the 1980s. Myxomatosis had originally been introduced in 1953 but had not survived (Filmer, 1953). The Parliamentary Commissioner for the Environment recommended against its introduction in 1987, largely as a result of public pressure (based on the perceived inhumaneness of myxomatosis). The Commissioner's recommendation was later endorsed by the government (Gibb and Williams, 1994). In 1991 yet another application was made to introduce myxomatosis and the flea vector. This application was again refused by the government in 1993, partly because of the promise of the potentially more humane biological control known as rabbit calicivirus disease, which was being tested in Australia (Williams and Munro, 1994). New Zealand then became involved in the Australian RCD research programme, something not widely known by New Zealanders at the time.

In rejecting the application to introduce myxomatosis in 1987, the New Zealand government recognised that hardship would be caused to farmers. The Rabbit and Land Management Programme (RLMP) was funded by government in 1989 for five years, to provide systematic and subsidised rabbit control to those farmers in the high country areas of the South Island who were hardest hit by rabbits. A total of NZ$16 million was set aside for the RLMP. In response to calls for more sustainable agriculture to be practised and for retirement from farming of the environmentally sensitive parts of the high country, the RLMP also included an extension campaign to advocate improved land management practices. The name of the programme was chosen carefully to reflect both aims. However, the two aims were not necessarily complementary, and it was possibly unrealistic to think that farm management attitudes and practices formed and reinforced for over 100 years could be changed in just five years.

The RLMP had a variable impact on rabbit numbers, and at many of the sites where rabbit numbers were monitored during the programme, rabbit counts did not reduce. At the cessation of the RLMP in 1995, the costs of rabbit control on the targeted properties again fell to the landholder. Up to that date, the continued subsidies had essentially postponed the day when farmers would have to pay the full costs of rabbit control on their properties. In terms of current political doctrines, this could be seen as maintaining the disempowerment and dependence of farmers, instead of gradually transferring the full responsibility and true costs. Alternatively, it could be seen as delaying the inevitable need to abandon or radically change farming practices on the vulnerable lands of the South Island high country.

As RCD testing progressed in Australia, the Rabbit Biocontrol Advisory Group (RBAG) was set up in New Zealand in July 1995, to advise government departments on issues relating to RCD. The membership of the RBAG comprised a diverse group of experts, scientists and interest groups, and it met several times until, with the imminent application to introduce RCD, it was disbanded in May 1996.

In addition to the 28 Australian native, feral and domestic animal species tested for their susceptibility to RCD, two New Zealand species were also tested. These were the lesser short-tailed bat (*Mystacina tuberculata*), one of New Zealand's only two native land mammals (the other being the greater short-tailed bat) and therefore the closest native animal taxonomically to the rabbit, and the North Island brown kiwi (*Apteryx australis mantelli*), one of several species of the national symbol. The kiwi had previously been the subject of much public concern during the debate over myxomatosis. When two kiwi reacted to RCD administered in the tests, widespread media publicity resulted and much public concern ensued. (It might be asked whether there would have been a similar reaction had the less appealing bat reacted to the tests, instead of the iconic flightless bird). RCD attracted further publicity when, in October 1995, it escaped from its test site on Wardang Island, 5 km off the coast of South Australia.

In December 1995, an RCD applicant group was formed by the Hawkes Bay, Canterbury, Otago and Southland regional councils, the Marlborough District Council, New Zealand Federated Farmers (Inc.), and the Office of the Commissioner for Crown Lands. Between them, the member bodies of the applicant group covered the likely beneficiaries of more effective and economical rabbit control, at least from a pastoral farming point of view. The applicant group and its consultants then prepared the application, which emerged in public in June 1996 as a substantial document of 229 pages, supported by 18 volumes of largely unedited technical appendices. In total, the application was 21 cm thick (RCD Applicant Group, 1996b).

The Ministry of Agriculture (MAF), the agency charged with making the decision about the introduction of RCD into New Zealand, adopted a decision-making process which involved more extensive consultation than was required by legislation. Public submissions were first requested on the criteria for making the decision about the introduction of RCD. MAF then called for submissions on the application itself, and received 751 responses. Several reports from experts in New Zealand and overseas were also requested. The decision, announced in July 1997, was a refusal to allow the introduction of the virus, on technical grounds. It was considered that not enough was known about how RCD worked, and that the biological control management programme that had been proposed was inadequate (MAF, 1997).

At this point farmers took matters into their own hands. In August 1997 the RCD virus was positively identified on several farms in some of the most rabbit-prone country of the South Island. Within days it appeared in other parts of the country, having been passed quietly from farmer to farmer. To date, nobody has admitted to importing the virus, and any assumption that it was imported illegally by a person is based on circumstantial evidence (and much rumour). However, several farmers have described how they helped spread the virus around once it was in the country (see, for example, *The Press,* 29 August 1997).

After briefly attempting to contain the spread of RCD, the MAF withdrew its efforts after only a few days. This approach contrasted with two previous successful attempts by MAF to maintain New Zealand's biosecurity, when in 1995 it eradicated an outbreak of fruit fly, and in 1996 an outbreak of tussock moth, both in the Auckland urban area.

The legal position of anyone involved in spreading RCD was unclear for several months. Spreading RCD appeared to be legal under the Biosecurity Act 1993, but illegal under the Animals Act 1967. Which of these Acts applied? The answer was 'both', and with the benefit of hindsight, this is clear from the title of the application itself (RCD Applicant Group, 1996b). Within days of the official discovery of the release, MAF declared that it would not prosecute anyone involved in spreading RCD, but would pursue vigorously anyone responsible for importing the virus. Despite this reassurance, regional councils decided not to become involved in spreading the virus without legal protection. Farmers, however, kept spreading it. They were initially protected from prosecution by the Biosecurity (Rabbit Calicivirus) Regulation 1997, which was gazetted in September 1997 under the Biosecurity Act. However, when it was discovered that such a regulation under the Biosecurity Act could not contravene the Animals Act, and was thus invalid, the regulation was replaced in March 1998 by retrospective legislation, the Biosecurity (Rabbit Calicivirus) Amendment Act 1998. A pure strain of RCD has been registered with the Pesticides Board, but it may still be illegal to spread the 'feral' strain of RCD for hire or reward.

What did the public think?

There is little mention of a role for the public in the brief history just outlined, but MAF did commission some research into public attitudes to rabbits and RCD in 1994, before it became a major issue (Fitzgerald *et al.*, 1994; summarised in Fitzgerald *et al.*, 1996). Public attitudes to rabbits and RCD have been described in detail in our research reports based on two series of focus groups and sample surveys of the New Zealand public (Fitzgerald *et al.*, 1996; Wilkinson and Fitzgerald, 1998), and the results will be only summarised here. The second of the surveys was carried out independently of MAF, with the aim of monitoring public attitudes during the debate on RCD.

Despite the publicity and information made available throughout the public discussion of the introduction of RCD to New Zealand, people's views

Table 14.1   Acceptability of current and potential rabbit control
             methods

Control method	Percent of respondents			
	Acceptable		Unacceptable	
	1994	1996	1994	1996
Shooting	83	76	9	11
Trapping	66	48	20	33
Imported natural rabbit-specific virus	39	42	41	37
Rabbit-specific genetically modified organism	48	42	32	36
Aerial use of 1080-poisoned bait	33	28	49	50
Use of other poisons (e.g. pindone)	28	23	41	39

Source: Fitzgerald et al. (1996); Wilkinson and Fitzgerald (1998).

of rabbits and the rabbit problem changed little between our two surveys, and also from the time of an earlier survey in 1991 (Sheppard and Urquhart, 1991). In both our surveys, more than 90 percent of the respondents thought rabbits caused environmental damage and a loss in farm production, yet only about half said rabbits were a concern to them.

As Table 14.1 shows, between 1994 and 1996, public acceptability of each of the current and potential rabbit control methods declined, with the exception of 'an imported naturally-occurring virus which is specific to rabbits' (a description of RCD which avoided mentioning it by name). Whether this reflects a heightened awareness of RCD following the Australian outbreak and the New Zealand debate, is hard to know. In our 1996 survey, 35 percent of the 600 respondents expressed unconditional support for RCD, with 12 percent giving conditional support. Unconditional rejection was expressed by 29 percent, with five percent giving conditional rejection and 16 percent taking an equivocal position (the remainder said they did not know). The main expressions of concern focussed on the question of whether RCD was specific to rabbits, whether enough research had been done on it and enough information provided, and whether enough was known about how it worked.

Another illustration of what the public thought about RCD is provided by the analysis of the submissions to MAF on the RCD application itself (Taylor Baines, 1996). As a method of assessing public opinion, the submissions certainly showed the range of opinions, but not the extent to which they were held. Those making submissions comprised a self-selecting sample which was heavily biased towards opposition to the application, was predominantly composed of South Islanders, and was polarised across all issues. Submissions also called for an independent panel to make the decision on the RCD application, rather than MAF. Under the new Hazardous Substances and New Organisms Act 1996, this is now the required procedure for such a decision.

Media coverage

The public debate about the introduction of RCD to New Zealand is best illustrated by the New Zealand print media coverage of the outbreak, following the unofficial release. According to our 1996 survey, 55 percent of respondents had seen information about RCD in newspapers, 77 percent in all media combined, and only two percent had obtained information anywhere other than in the media. The coverage was a typical media mélange of 'whodunnit' sensationalism, the reinforcement of stereotypes, and straight factual reporting. While in the first few days of the outbreak the main theme of the stories was the power relationship between farmers and MAF, the longest-lived theme was confusion over the legality of spreading RCD.

Outbreak

On day 1 the news media reported the known facts about the release of RCD, and reactions to this ('Anger, joy as rabbit virus found', *The Press*, 27 August 1997). On day 2 the signs of the Government's acceptance of the introduction of RCD and the inevitability of its spread were already apparent ('Govt may exploit rabbit virus', *The Press*, 28 August 1997). By day 3 reports indicated that it was too late to stop RCD, as farmers had already spread the virus throughout much of New Zealand, and were telling stunned officials that efforts to contain it were futile ('Farmers spread virus NZ-wide', *The Press*, 29 August 1997). Reports on day 4 indicated that the fight to eradicate or even contain RCD was clearly over, with attempts to quarantine the disease abandoned ('Govt likely to abandon RCD fight', *The Press*, 30 August 1997). It was already evident that the virus had arrived in New Zealand, and that farmers had orchestrated its spread with precision.

Legalities

The legal position of anyone involved in spreading RCD was not clarified for some time, and appeared to fluctuate with the discovery of each new piece of relevant legislation and each new legal opinion (Figure 14.1). Initially farmers were reassured that only the importation of RCD was illegal, and that farmers who spread it once it was in New Zealand would not be prosecuted under the Biosecurity Act ('No prosecution over RCD confessions', *The Press*, 29 August 1997). A week later though, a new legal opinion from the Crown Law Office indicated that anyone caught spreading RCD could be prosecuted under the Animals Act ('Spreading virus breaking the law,' *The Press*, 6 September 1997). The government then decided to regulate under the Biosecurity Act to permit the spread of RCD, and recognised that MAF's initial reassurances left it in no position to prosecute farmers who admitted spreading the virus ('Little fear of prosecution over RCD', *The Press*, 9 September 1997). This regulation was eventually gazetted almost a month after the outbreak ('RCD made legal', *The Press*, 25 September 1997). When the regulation was found to be invalid ('RCD regulations illegal: committee', *The Press*, 24 February

1998), legislation was required ('Rabbit virus legal', *The Press*, 8 April 1998). As RCD became older news, these stories retreated from the front pages, and eventually became one-paragraph items only.

The players

The power wielded by the farmers was clearly demonstrated ('They wouldn't tell if they did know', *The Press*, 28 August 1997). Along with evidence of the farmers' power, media stories showed sympathy for their plight, with farmers suggesting that official procrastination on the rabbit problem and the costs of control had finally forced someone's hand. Whether or not the farmers shown in Figure 14.2 knew how RCD arrived in New Zealand, they still managed to be portrayed sympathetically. The clearest indication that farmers had taken charge and that MAF was powerless came from the debate among farmers about whether they would share their knowledge of RCD's effects and use with MAF, or whether they would proceed alone ('Farmers urged to share RCD data with MAF', *The Press*, 25 September 1997).

Farmers showed considerable ingenuity in spreading RCD. For example, in the very early days of the outbreak, and without the knowledge of council staff, farmers spiked with RCD carrots which had been stockpiled for use in a Canterbury Regional Council rabbit poisoning campaign ('Farmers spike council carrots', *The Press*, 30 August 1997). Farmers also explained how they had conducted methodical trials to determine the best formulation of virus mixture and how best to spread it, even mixing so-called 'rabbit smoothies' in a home blender, using organs from infected rabbits ('Rabbit smoothies help spread virus', *The Press*, 30 August 1997).

With the media images of 'rabbit smoothies', 'kitchen whizz' strains of RCD, white-coated scientists testing for viruses, and farmers holding up dead rabbits, the public received plenty of entertainment and reinforcement of stereotypes of farmers and officials alike. However, it was a fortnight before a measured explanation of what was and was not known about RCD and its behaviour was published in the newspapers ('RCD: clearing the confusion', *The Press*, 11 September 1997).

Rural media

By the end of the second week after the official discovery of the release, detailed coverage of RCD issues had become confined to the specialist rural media, in which mixed messages continued to be given. The free, nationally

Figure 14.1 Spreading RCD: Legal or illegal?

Figure 14.2 Who had the power? MAF? Farmers!

distributed *Rural News* reported that, despite the gazetting of the regulation to make the possession and spread of RCD legal, some officials in several agencies thought the spread of RCD was still not entirely legal ('RCD chaos', *Rural News*, 6 October 1997). In the same article, unnamed 'biosecurity experts' were reported to be critical of 'backyard dabbling by do-it-yourself kitchen whizz 'biochemists' in the South Island'. Adding to the confusion, a North Island pest control contractor was also pictured with his gun and his dogs, saying he was ready to spread RCD for farmers at NZ$50 per hour ('RCD spreader for hire', *Rural News*, 6 October 1997). At the same time, *Straight Furrow*, the official organ of Federated Farmers of New Zealand (and distributed free of charge to all New Zealand farmers), gave farmers some tips ('Getting the best from your RCD', *Straight Furrow*, 13 October 1997).

Persistent views

The confusion over the legality of deliberately spreading RCD led to differences between regional councils over the assistance that could be given to farmers. Some councils, such as Marlborough, opted to assist farmers to spread the virus as much as they could without risking prosecution ('Council helps farmers to spread RCD', *The Press*, 29 November 1997). However, regional councils in the North Island, citing continued confusion over the legal issues, elected to leave the spreading of RCD to farmers. Questions over the legality of spreading RCD persisted, and every few weeks a paragraph with another legal twist appeared in the newspapers, until the retrospective legislation clarified the issue.

Even after the initial wave of interest subsided, news about RCD was still occasionally splashed across the pages of the major newspapers. For example, in late November 1997, the *New Zealand Herald*, the major Auckland daily, carried a feature which, despite the sensation-seeking headline and the accompanying photograph of four dead rabbits strung on a fence, was generally sympathetic to the plight of farmers looking for a way to control rabbit numbers ('The man who killed Peter Rabbit', *New Zealand Herald*, 29 November 1997).

Issues and reflections

Victim of circumstance

As the media coverage shows, the public debate about the introduction of RCD to New Zealand was not systematic, having twice been overtaken by circumstance. As a result it escalated too quickly for the public to be able to keep up with the issues and stay informed. The proponents of RCD found themselves attempting to respond to key issues at short notice, while their opponents were provided with opportunities to muster public concern about risks and vulnerabilities.

Essentially, the chance of systematic and timely debate about the use of RCD was lost with the outbreak of the disease in Australia. The attendant public and media reactions initially centred on the unknowns about the disease, including how it was transmitted. Debate then moved to, and became embroiled in, the issue of the kiwi. The emotions associated with the consideration of the potential for the kiwi's destruction took root before the public had a chance to learn about the issue in a structured way, and give systematic consideration to RCD as a form of control for the rabbit in New Zealand. Fear therefore characterised the debate from its earliest days.

Public information

Even when the opportunity for public debate existed, officials did not actively manage the situation. While a draft communication strategy on RCD was developed by the government (Wall, 1995), there is no evidence that official information reached the public to any significant extent. The volume, technical inaccessibility, and limited availability of official information meant that the public had little opportunity to become fully informed. Official information provided to the general public consisted of an information kit produced by the government's Rabbit Biocontrol Advisory Group (RBAG, 1996). Documentation accompanying the application to import and release RCD was prepared with the government's requirements in mind, not the public's need for information, although the applicant group also produced a public information kit (RCD Applicant Group, 1996a). However, none of these documents reached widely into the community, and none of the members of the public who participated in our focus groups had seen them until we displayed them to these groups. In our 1996 survey, 63 percent of respondents disagreed or strongly disagreed with the proposition that there was sufficient information available to the public about RCD.

Public involvement

In our original research report to MAF we recommended, on the basis of our first round of research on public attitudes to biocontrol, that genuine public and stakeholder participation would need to be fostered (Fitzgerald *et al.*, 1994: 77). We further recommended, in our more widely-distributed summary report, that an active programme of public involvement would be required if a decision acceptable to the public were to be made (Fitzgerald *et al.*, 1996: 20). We warned that communication with the public was difficult and risky, quoting the comment by Morgan *et al.*, (1992: 2055) that 'one should no more release an untested communication that an untested product'.

It is worth reiterating the point made earlier, that the official decision-making process itself involved more extensive consultation than required by the existing legislation. MAF initially sought public submissions on the decision criteria to be used when considering an application for importing and releasing RCD. Later they sought submissions on the application itself. This was, however, a standard institutional approach to public participation in decision making. The extent of public information, consultation, and involvement was minimal, given the potential for controversy.

Public cynicism

When farmers were found to be spreading RCD around New Zealand within weeks of MAF's decision not to allow its release, the public was left to conclude that farmers put their own views and interests ahead of all others, had no regard for due process, and were willing to take the law into their own hands. Further, MAF's low-key reaction implied that farmers would be allowed to act this way. Public discussion has subsequently broadened to include not only the nature, properties and risks of RCD, but also the futility of public debate and participation in decision making, given the ease with which micro-organisms (or any other organisms) can be introduced to New Zealand, either illegally or accidentally. Another debate has emerged about what to do about a group of people, such as farmers, who flout the law.

Despite statements by government ministers and MAF officials that whoever was responsible for introducing RCD to New Zealand would be prosecuted, no such prosecution has yet been brought, and no one has been prosecuted for spreading the virus. The question remains - will anyone ever be prosecuted, given the legal confusions and given that the main victims of the virus appear to be the unwanted rabbit, the nascent rabbit meat industry, and the authority of MAF?

Public cynicism has possibly been exacerbated by the apparent conflicts of interest of MAF. Having provided financial support for the testing of RCD in Australia, MAF was then obliged by its own processes firstly, to decide that it was not yet time to release the disease in New Zealand, and secondly, to try to control the outbreak when it did occur. Fortunately, the new Environmental Risk Management Authority, charged with decision making about the introduction of new organisms under the Hazardous Substances and New Organisms Act 1996, will not have to face such conflicts.

Farmers' concerns

When the decision not to release myxomatosis was announced in 1987, it was accompanied by the announcement of a publicly-funded rabbit control programme, the Rabbit and Land Management Programme, which recognised the plight of farmers at the time. However, the decision not to release RCD was not accompanied by any similar recognition, and no offers of even temporary relief were forthcoming from government. The farmers believed they were not being heard in the RCD and rabbit control debate.

Conclusion

Despite public opinion and concerns about RCD (as measured in our surveys and voiced in the submissions), and despite MAF's decision to not introduce RCD on the basis of inadequate technical understanding of the disease, nevertheless RCD has been introduced into New Zealand. That this happened is a reminder of the limitations of legislation and of traditional conservative power structures in the face of a strong determination to disregard the law. Such limitations are especially evident in cases such as this, where micro-organisms are involved.

Finally, there still remains a strong need for a public education campaign about RCD, in order to provide answers to some basic questions. The only public education that has been achieved to date seems to be on the need to vaccinate pet rabbits against the disease. After the unofficial release of RCD, there is still a need for better and more effective public information.

References

Filmer, J. (1953), 'Disappointing Tests of Myxomatosis as Rabbit Control', *New Zealand Journal of Agriculture*, Vol. 87, pp. 402-404.

Fitzgerald, G., Saunders, L. and Wilkinson, R. (1994), *Doing Good, Doing Harm: Public Perceptions and Issues in the Biological Control of Possums and Rabbits*, New Zealand Institute for Social Research and Development: Christchurch.

Fitzgerald, G., Saunders, L. and Wilkinson, R. (1996), *Public Attitudes to the Biological Control of Rabbits in New Zealand*, MAF Policy Technical Paper 96/3, Ministry of Agriculture: Wellington.

Gibb, J. and Williams, M. (1994), 'The Rabbit in New Zealand', in Thompson, H. and King, C. (eds), *The European Rabbit: The History and Biology of a Successful Colonizer*, Oxford University Press: Oxford, pp. 158-204.

Ministry of Agriculture (1997), *Decision on the Application to Approve the Importation of Rabbit Calicivirus Biological Control Agent for Feral Rabbits*, Ministry of Agriculture, Wellington.

Morgan, M., Fischhoff, B., Bostrom, A., Lave, L. and Atman, C. (1992), 'Communicating Risk to the Public: First Learn What People Know and Believe', *Environmental Science and Technology*, Vol. 26, pp. 2048-2056.

New Zealand Herald, Auckland.

Rabbit Biocontrol Advisory Group (1996), *Rabbit Calicivirus Disease Information Kit*, Dunedin.

RCD Applicant Group (1996a), *Rabbit Calicivirus Disease: a New Opportunity for Rabbit Control in New Zealand*: Dunedin.

RCD Applicant Group (1996b), *Import Impact Assessment and Application to the Director General of Agriculture to Approve the Importation of Rabbit Calicivirus under the Animals Act 1967 and to Issue an Import Health Standard under the Biosecurity Act 1993*: Dunedin.

Rural News, Auckland.

Sheppard, R. and Urquhart, L. (1991), *Attitudes to Pests and Pest Control Methods*, Research Report 210, Agribusiness and Economics Research Unit, Lincoln University: Lincoln.

Straight Furrow, Wellington.

Taylor Baines and Associates (1996), *Analysis of Submissions on the Importation Impact Assessment for the RCD Virus*, Ministry of Agriculture: Wellington.

The Press, Christchurch.

Wall, M. (1995), *Draft Communications Strategy for a Public Information/Communications Programme on Rabbit Calicivirus Disease (RCD)*, Ministry of Agriculture: Christchurch.

Wilkinson, R. and Fitzgerald, G. (1998), *Public Attitudes to Rabbit Calicivirus Disease in New Zealand*, Landcare Research Science Series No. 20, Manaaki Whenua Press: Lincoln.

Williams, R. and Munro, R. (1994), 'Community Attitudes', in Munro, R. and Williams, R. (eds), *Rabbit Haemorrhagic Desease: Issues in Assessment for Biological Control*, Bureau of Resource Sciences: Canberra, pp. 71-78.

15 Hiring labour for sugar harvesting: Farmers, farm workers and sub-contractors

MICHAEL FINEMORE AND JIM McALLISTER

Introduction

In a recent article on contemporary Australian agri-food restructuring, Lawrence (1996: 60) proposed that:

> The beef feedlot...changes the nature of 'agricultural employment': the grazier producing extensive beef gives way to the unskilled worker in the feedlot. Those graziers remaining - and who provide either the grain or the store animals for the feedlots - appear to have reduced autonomy and flexibility.

We do not quarrel with this proposal, since it seems to us to be one of the likely outcomes of agricultural adjustment - not just for beef, but for a number of agricultural commodities. But our recent experience of field research suggests at least one alternative scenario. This paper investigates the implications of the apparent move by farmers to protect their central investment in their farms by excising parts of the production/labour process (and letting them out to sub-contractors), and by withdrawing towards a core of activities which can be reproduced as 'family-farm/owner-operatorship'. The nature of these core activities depends on the structure of farm enterprise, the preferences of the farmer, the specialisations of members of the farm business, and so on. Our interest is not in discussing precisely how the macro-restructuring issue is accomplished, nor indeed with the local concern about what parts of the family farm might be considered 'central'. This chapter concentrates instead on 'the third ingredient' - farm work, and the social consequences for farm employees of changes in the structure of the Australian sugar industry.

The study of paid agricultural labour

Rural sociology in Australia has been principally concerned with three issues. The first of these is the place of family farming within capitalism (Lawrence, 1987; Gray *et al.*, 1993); the second is concerned with issues of land and water degradation and of sustainability (Lockie and Vanclay, 1997; Vanclay and Lawrence, 1995); the third concentrates on 'country life' - not just in terms of agriculture, but in terms of towns as farm service centres and seats of farmer power. The latter includes the form and extent of rural social disadvantage and poverty.

Rural sociology in Australia is also noteworthy for a number of significant exclusions - although these are gradually being overcome. The first is the role of women (see James, 1989; Alston, 1992); the second is the concern with agricultural commodities as human food (see Parsons, 1995); the third (and the one which we analyse in this chapter) is the place of paid agricultural employees in agricultural production. This latter issue has been addressed in a number of historical studies, and in the study of industrial relations in agriculture - but not in sociology. There has been some analysis of the place of Pacific Islanders in agricultural production in Queensland in the work of Gistitin (1995) and the Evatt Foundation (1991), and the role that Aborigines have taken in rural industry is discussed by Rowley (1972). Thorpe (1992) has formalised this concern in creating his concept of 'colonised labour'. Taylor (1997; see also May, 1994) has recently emphasised the degree to which the grazing industry in Australia was able to gain a competitive advantage by dispossessing the Aboriginal people of their land and thus making land a free 'factor of production'. Encel (1970) has described graziers and their contribution to Australian society, both as agricultural producers and decision makers throughout the country, and in terms of their role as a social elite. Gill (1981) has analysed farm employees, comparing the difference in lifestyle, work habits and patterns of deference of workers on grazing properties as compared with the itinerant shearers. Finally, Ruben (1992) has critically analysed the conceptualisation of farm-family agriculturalists which characterises them as independent owner operators, when in fact they are supplemented on an annual basis by an itinerant labour force which harvests the fruit. In other words, these family farmers are enriched by this opportunity to 'exploit' the labour of agricultural employees.

Contract employment in agriculture

The literature considered thus far, proposes three types of paid farm workers:

- 'industrial agriculture' employees (including intensive livestock workers);
- 'harvest labour' (usually itinerant), and;
- on-farm employees ('farm hands').

Recent investigation of the first category (Lawrence, 1996; Lyons, 1996) appears to show two labour processes - that producing 'the unskilled worker of the feedlot', and that involving a contract farmer. The former mirrors the 'factory' employee; the latter is a 'propertied labourer' (Davis, 1980; see also Miller, 1996), renting farm facilities and selling family labour to a corporate entity, such as a processor or supermarket (Rickson and Burch, 1996).

Within the fruit picking context, the implicit work agreement that a 'harvest labourer' has with the farmer is referred to as 'a contract', implying that a certain amount of work will be done at an agreed-upon piece-rate of pay. Even as a verbal agreement, this is hedged around with 'understandings', for example, that there will be flexibility with regard to maintenance of a level of care in fruit handling, met on the employer's side by a willingness to pay 'above agreement' rates for difficult-to-harvest sections of the crop.

At one level, the farmer can be seen to have separated one section of the crop production process and ceded its control to 'a contractor'. The logic of divisibility of the process into discrete stages, each of which can, in theory, be delegated to a new contractor, parallels the Taylorist notion of industrial efficiency. In citrus growing, tree-planting, irrigation, pruning, pest and disease control, harvesting and processing and/or marketing, are production segments which are sometimes contracted out. While this apparently resembles the approach of Goodman *et al.* (1987) in which the stages of production are appropriated and then sold back to the farmer, as well as some of the variations of *disconnectedness* which van der Ploeg (1992) explores, in fact the contracting process as formulated here differs from both these approaches.

When the task of harvesting has become too big and too complex to be handled efficiently by a verbal agreement between farmer and picker, the need for a 'contractor' is, within the Taylorist logic, obvious enough. Within the fruit and vegetable industries of California this relationship has developed in a context of (international) migrant labourers and 'agents' (Mooney and Majka, 1995). It has an historical resonance in Australia and New Zealand in contracted shearing teams, and in Australia in an earlier era, with sugar-cane cutting gangs:

> Canecutters generally worked under a piecework system, i.e. they contracted to cut a set number of ton[ne]s in a season and were paid at a fixed rate for each ton[ne] cut. In the far northern region...a gang of up to eight men would sign a contract with one or more farmers in the one area to cut an agreed tonnage. The earnings would be shared equally amongst members (Balazategui, 1990: 78).

During the manual harvesting era, canecutters were employed for three major tasks: burning, cutting, and loading.[1] This harvest labour force consisted principally of men aged 21-35 years, and a high proportion were itinerants. Their work was arduous but their earnings were relatively high (Anon, 1970: 31).

> Historically...cane was burnt...to lessen the incidence of Weil's disease. Furthermore 'it cleaned the trash and weeds, got rid of snakes...made the cutting...easier and complied with...mill demands to have clean cane for crushing'. Burning became universal practice [only] during the [Second World] War because the speedier harvesting that resulted, compensated slightly for the paucity of labour (Balazategui, 1990: 60).

Burnt cane stalks remained undamaged, although that burning marginally lowered the sugar content of cane. As a consequence burnt cane (even cane accidentally consumed by wild fire) had to be harvested fast to maintain quality:

> [New gangs were usually] shown how to burn cane the evening before [they] commenced work and...taught how to cut and load cane by the farmer on the first day of cutting (Balazategui, 1990: 61).

One of the most profound changes impacting on the sugar industry and the labour force was the introduction of mechanical harvesting. Moreover, as these machines became ever more efficient, they replaced an ever-larger proportion of the manual labour force until,

> [by] December 1977 all reference to manual canecutters was deleted from the Sugar Industry Award (Balazategui, 1990: 97; See also Hungerford, 1996: 42-56).

Today, Drummond (1996: 10) notes the

> almost total commitment to modernisation which has been a key feature of
> the Australian industry...at the forefront of the development of specialised
> technology.

This transformation has had a significant impact on both the sugar industry
and the wider communities associated with the industry. In order to understand
the extent and the nature of the changes in sugar cane harvesting, and the
impact of these changes upon labour, it is necessary to contextualise the
industry, both sociologically and historically, and to provide a backdrop against
which more recent changes to industry work practices can be understood.

Articulation of the sugar industry with the coastal towns

When cane was cut by itinerant teams, maturity of the annual crop had two
obvious local consequences: the annual influx of workers made heavy demands
on services, and the slower tempo of social life quickened (especially on
weekends). This greater activity had both positive and negative outcomes.
Demands on accommodation, goods and services, and on public and civic
services, all increased, and the towns became bustling mini-metropolises. At
the same time, road traffic hazards increased, as did civil disturbances, in
streets, pubs and dance halls and air pollution increased markedly (once cane
burning became accepted practice, just before the Second World War).

But the season lasted only from early in June until Christmas. During
the non-cutting season (or 'the slack'), cutters had a choice:

> If cutters decided to stay in the district a number would find fieldwork: hoeing,
> cutting plant cane, feeding a planter, for some part of the slack. If they remained
> in the barracks the farmer might recruit them to pull out trees in uncleared
> land or repair and paint the barracks. A popular slack period destination...was
> the tobacco region of Mareeba, to which many cutters, particularly migrants,
> returned slack after slack just as they journeyed back to the sugar towns season
> after season...Migrant canecutters often tried a wide range of jobs in the slack
> season. One who claimed to have tried virtually every job available had been
> ringbarking, constructing dams, working for the Main Roads Department,
> labouring on the Cooktown-Laura power line, working in meatworks, and a
> chocolate factory, and driving trucks (Balazategui, 1990: 74).

Alternatively, a worker might return south, either to work there in possibly more congenial circumstances, or:

> go to Sydney and 'do nothing but sleep' - to hibernate right through the slack...[S]ingle, relishing freedom of opportunity, the sense of safety and plenty, but bone-weary after the brutal rigour of the season, [cutters - Roo and his mates, from *The Summer of the Seventeenth Doll* (Lawler, 1959), for example] longed for rest and aimless distraction (Balazategui, 1990: 75).

What occurred when cutters remained in town? Some workers wanted the security and prosperity of local full-time work; and businesses liked the idea of year-round economic activity occasioned by increased population living and working in the regions; and farmers wanted an assured labour supply to begin work at the start of the next season.

This latter factor was especially important, and the political manoeuvring by the farmer organisations and the Federal Department of Labour and Industry on the one hand, and the Australian Workers Union on the other demonstrates the level of insecurity engendered in the farmers by seasonality and the lengths they would go to increase their control over their labour supply, and thus over the whole labour process (see Balazategui, 1990). Of course, on many farms there was a 'hired hand' or two tending the non-harvest aspects of the crop cycle: ploughing, planting, fertilising and irrigating. Their incomes were much lower than those of cutters, but some were prepared to accept relatively poorly-paid jobs during 'the slack', believing that the level of their wages during the cutting season 'averaged out' (Balazategui, 1990).

The impacts of mechanisation

The mechanisation of sugar cane harvesting had major impacts at all points of the production process. In this survey, we can only focus on issues relating to labour and the conditions under which it is supplied, and the contract harvesting process.

The conditions of labour

The consideration of the supply of labour, especially wage-labour, to the sugar labour process has to deal with price and availability of working people. Base rates of pay for seasonal work are set higher than base farmhand rates to allow for an element of the seasonality of this work, but the fact that seasonal workers are attracted to 'overtime' must suggest that their normal working routines eat into their potential leisure.

When the season is over and seasonal workers experience economic recession, they can leave the district to pursue other seasonal work, stretch their earnings over a year, or try to find work in their local town. Compared with the era recounted by Balazategui (1990), much of the industry of local towns has now been consolidated in the key coastal centres or has moved 'offshore'.

The Bundaberg region of Central Queensland, for example, has a reputation for having Australia's highest unemployment rates for some categories of workers. Pay rates are also relatively low. The 'award' rate is established by negotiation between Canegrowers, the industrial organisation of the sugar cane producers, and the Australian Workers' Union. The relatively low remuneration for farm work is compounded by the controls which operate on local labour to keep their industrial demands 'moderate'. On two continents, Newby (1974) and Barlett (1986) have drawn attention to the docility of 'farm hands' (a product of their 'special' industrial relationship with their employer), and our interviews reveal too, that a local labour force is considered to be more manageable than one composed of outsiders. Recent innovations in the milling side of harvest - continuous crush and extended harvesting hours (McAllister *et al.*, 1995) - have been predominantly intended to improve machinery (and hence economic) efficiency, but the most recently mooted innovation - to extend the harvesting season - can be interpreted as a manoeuvre to maintain the harvest labour force in steady work within the region. In the language of the growers the 'slack' still exists; whether it has any reality for harvest workers is not so clear.

What Newby (1974) has called 'deference' among British agricultural employees seems to be experienced by the working class in Australia as 'country-mindedness' - acceptance of the dominant culture resulting from a subtle blending of cultural hegemony and implied force. Even in sizeable country towns, farm owner-operators hold social pre-eminence (Higgins, 1996), but 'reputation' - as a 'good quiet worker' and someone who 'stays out of trouble' - puts job-seekers nearer the front of the hiring queue.

Contracting of sugar harvesting

The timeliness of harvesting operations is critical in the sugar industry. The commercial cane sugar content (CCS percent) rises as the crop matures. CCS deteriorates (through fermentation) the longer cut cane is left in trucks unmilled, and thus the shortening of the time from 'cut to crush' also maximises returns to growers. Efficiency of the transport system, governed by centrality of the mill within its mill area and the ability of the mill to receiving cane at all hours (McAllister *et al.*, 1995) also contributes to crop profitability. Since payment is a product of volume of cane milled and its CCS, farmers prefer to harvest all their cane during the very short period when sugar content is at its peak. However, the quantities of cane and mill capacity make this impossible, and mills have traditionally enforced a 'roster' system of harvesting. Each farm is permitted to harvest and mill a proportion of its crop, to ensure both fair treatment of all mill suppliers and a coordinated supply of cane for crushing. The principal inspector of each mill has the task of coordinating the harvest over the wide mill area. The roster system introduces potential conflict between growers and inspectors in relation to which crops are harvested and at what time; maintaining round-the-clock fresh cane to the mill and ensuring that farmers do, indeed, cut a considerable proportion of their cane when it is nearest its maximum CCS, reduces this tension somewhat. Traditionally, cooperative mill ownership has given farmers a measure of control over the equity with which their crop is 'manufactured'.

Family farming is a feature of cane production in the whole Australian industry, but just what that expression means seems to be changing quite rapidly (Djurfeldt, 1996). Nevertheless, 'plantation' production on land owned and operated by the milling companies is making a comeback, and deserves further study. Many farmers harvest their own cane; some farmers harvest as 'groups'; and contract harvesting is a feature of all districts now, probably due both to the inability or reluctance of growers individually to purchase and operate such expensive equipment, and to the relatively small property size. The 'contractor' in cane harvesting, and the similarity of that function to the 'shearing contractor', has been mentioned already, but the cane harvest contractor is usually involved in the harvest itself. Producers in Mackay, North Queensland, may characterise contractors as merely 'bright young entrepreneurs', but they are usually farmers' sons and/or farmer/contractors. In other words, contractors may be farmers (with varying sized enterprises) supplementing or diversifying their income through contracts with neighbours; 'second sons' (sic) of farmers, financed by a lien on the parental property; or independent business people seeing the possibility of a livelihood; and so on.

As suggested above, only a small proportion of contract harvesters operate in that role full-time. Investments of A$100,000 to A$250,000 in tools-of-trade and/or mobile property (to cut 80-100 thousand tonnes per season), identifies these contractors as part of the 'new rural middle class', despite the indebtedness which most claim attaches to their property. With such indebtedness, they are under considerable pressure to improve their efficiency of machinery use, which implies increased hours of utilisation of ever-larger machines and the introduction of long working hours (often into two eight- or nine-hour shifts) through expanding the size of enterprise. In this respect, contract harvesters are highly regarded by the millers, who also want to extend hours per week of harvesting, again chiefly for economic reasons. The contractors' employers are individuals and/or groups of farmers usually located close together; the contractors' competitors are independent farmers, and other contractors.

An important aspect of the contracting system is that the size of machinery and the mode of its use prevents contractors working with individual smaller farmers and on isolated hill farms. As one interviewee noted:

> [S]ome of those fellows in their rougher places are cutting their own cane because they want to, but some of them are cutting their own cane because they have to. No-one, no contractor, is going to; you simply can't get the through-put.

Small isolated farms continue to harvest their own cane then, or turn to cooperative arrangements among neighbours for cash payment or work exchange. When they do their own harvest, it is often with outdated machinery. For farmers who also contract to cut their neighbour's crops (occasionally up to 30 kms away), harvest costs are increased by travel time, but parties to such a business arrangement may at least be able to rotate farm activities (replanting, irrigation and fertilisation, etc.) with their harvest responsibilities. Moreover,

> Some people don't like contractors because...they go through too fast, or they don't cut it the way you want to do it, or it's part of your job and you like to do it: 'that's my life. My life's growing cane, and I should be able to have that right'.

Business competition is fierce. On one hand, larger contractors have debts and other payments to make, while on the other, the competition among contractors forces down the price of contracts so that the profit margin is trimmed. In the middle, transport, maintenance and labour costs eat into the business returns. Labour costs involve remuneration for work driving tractors, harvesters and haulouts.

According to interview data, in order to maintain their labour force, both farmers and contractors pay a base pay rate plus 'skill loadings', and then offer overtime to make this seasonal work attractive:

> There is a general view that you wouldn't [be able to] find good people or you'd lose them if you didn't give them all the overtime....

Many Mackay employers remarked about the value of a good worker to their enterprise:

> You've got them working half the night and most of the day....

> If the man doesn't turn up at 5 o'clock in the morning, you're down a man for a day. You just can't bypass him or pick a man up.

> If a man falls out, our production goes down by a half (or more) for the day.

And, perversely,

> You pay their superannuation, their workers compensation, give them their 'top dollar' for wages and overtime.

Nevertheless, it was equally evident that they resented what they characterised as 'excessive' wage demands:

> In the structure of the system at the moment, there is flexibility, but not as flexible as I'd like to see it.

The bigger contractors were constantly searching for means to increase their production without proportionate increases in their labour force; considerable ingenuity was expended planning the labour process to keep labour costs down, with one farmer proposing:

> It's going to come to a stage eventually where you've done your 10 hours work and that's where you say 'Look, I can pick up [a contract for an extra] 10,000 tonne because this group's splitting up'. [But] I can't really work my men any more than 10 hours, do I go to a seven day roster instead of five? And when that's full up, say 'look, that's it - the only way I can improve my efficiency is to go to the two shifts!' I believe the two shifts is the next step - four [am] to twelve [midday] and twelve [midday] to eight [pm]....

Another said:

> After, say, 10 hours a day for four days, you've got one crew; they've got their 40 hours, and you move another new crew in for another 10 hours for the next four days.

And a third proposed:

> I...looked at a system where you had three carters and a driver, but you actually put five blokes on - so, every two days you're paying five men to cut. If you say 70,000 tonne, you put five men on; every day there's a man off: so he has his two days off, he comes back, the next bloke goes off. So, instead of putting two shifts on, you actually pay to put the one extra man on.

Consequences of change

Techno-economic change within the sugar industry has also brought social costs. As one Mackay farmer reminisced:

> Well, my dad and myself - back in the cane cutting days - of a Friday afternoon at dinner time, you put your cane knife on the post and you went home. They've still got that mentality, you know. Look at that hullaballoo when 'continuous crushing' come in. Everyone said, 'it will never work - we'll never bloody do that. We're not working Saturdays - we're not doing that!' They all done it; there's benefits in it.

Farmers certainly have borne part of these social costs but, as always, costs fall differentially upon people involved in change, and contractors and paid workers are not exempted, either. As mills in various seasons have extended their work week to a seven-day 'continuous crush', farmers' work (i.e. harvesting) has come to spread over seven days a week, for 22 weeks, on a rostered basis. Views on sharing the work differ somewhat. As farmers tell it, they are 'working every hour God gave'; in the recounting of a contractor:

> I cut seven farms - they don't work Saturday and Sunday...You're the only bloke out there doing it.

Almost invariably, though, paid workers are out there too, and the fractured nature of their work-lives stamps its disruption on their social and family lives as well.

Full time 'farm hands' are still maintained on some farms. However, many farms have reverted to using only family and season labour (Hungerford, 1996: 184). Meanwhile, the conditions of work for sedentary farm employees are changing as mills make 'untimely' delivery demands on farm employers. Equally, harvest workers appear to be adjusting to the change from single-farm employment, to employment by contractor, which implies an increase in shift work to service mill demands. A few farm hands may move from general farm work to harvest labour in season, but most harvest labourers are designated as temporary or seasonal labour. Harvester driving is somewhat more skilled than transporting the produce, but both jobs involve control of expensive and specialised mechanical equipment. Farmers and contractors rightly believe that they should allocate such tasks only to workers they know and trust. Considering the costs associated with harvest labour, some farmers preferred to keep a 'farm hand' to do the routine work and handle the harvest themselves.

Paid work teams are occasionally related by kinship to the farmer or contractor but employers select their workers by personal knowledge. Employers work to keep good staff, not just through the 22-week harvest but season after season. Some farmers do off-farm work, and undoubtedly some of the seasonal workers are 'pluriactive' farmers who complete their own farm work when not working for wages. But some, at least, represent the other end of the economic scale and have been,

> lucky enough to pick a couple of farms up close to [the home property] and run [an extended family enterprise].

Conclusion

Drummond (1996: 13-14) sees the current processes of structural adjustment as follows: although social conflict has been avoided, postponed and/or defused in the two eras of capitalist regulation in this century up to the present, we have entered a crisis period now, in which the existing mode of social regulation has broken down and the relationship between capital and labour needs to be renegotiated if private capital accumulation is to continue. For us, farm ownership seems to imply that farmers have decision-making authority and the power to control their own work, that of their employees, and the way in which the farm is managed. If growers are separated from control over their own decision-making in relation to harvesting, in what ways do they attempt to regain that control? One might consider, by contrast, what decision-making prerogatives reside with the contractors. In other words, if contract harvesters are assuming increased control over decision-making relating to harvesting, where does this leave the grower? Is there, for instance, a consequent loss of grower autonomy, or is contracting also a grower strategy?

Answers to these questions are not available yet for any of the agricultural commodities requiring a harvest labour force, although farmers who make contracts in this sense, seem to be much better positioned to dictate at least some of the conditions of their production process than are those who take contracts from oligopolistic companies. The implication here is that, depending on the roles taken, a contradictory class position (Wright, 1978) may develop, for instance, in the case of a grower hired to harvest other growers' crops but employed by a non-farm-owning contract harvester.

How then do these people, confronted by social change, reconcile these contradictory identities while at the same time attempting to reproduce their means of existence as growers, in order to maintain farming as a way of life? In light of the particular development in restructuring considered here, we are prompted to ponder: is paid work viable or is it being superseded? For the present, at least, contract harvesters do not seem also to subcontract their 'workers' - to engage in 'profit sharing' rather than paying wages. Were they to do so, 'workers' would then become, in effect, business associates rather than employees. If this was the wave of the future, 'employers' would be in a position to drastically undercut current regulation of 'surplus' sharing, and to introduce the conditions for a much more drastic exploitation than exists already of the harvest labour force in sugar (and probably other) production systems.

Peoples' situations in the struggle to gain access to the means to a prosperous livelihood depend on their stake in the production process and their combination of work activities. Wells (1984; 1996) drew attention to the case of strawberry production in California where (in regulationist terms) capitalist producers of a niche crop invited their harvest labourers to participate in the 'ownership' of the crop, by 'share farming' its production, rather than just harvesting it. However, these new 'growers' found themselves in exactly the position Rickson and Burch (1996) describe for farm owners who become contractors - they carry all the risk and indulge in 'self-exploitation' throughout the production cycle. The original argument about two forms of contracting seems to have brought us full circle, back to considerations of subsumption. In brief, instead of going into debt for their harvesting equipment - like Lyons' (1996) poultry farmers, borrowing to upgrade battery cages - some sugar farmers shift indebtedness to another 'species' of independent owner-operators and, in the process, create the class of contractors (Smailes 1996: 317-319). As a consequence,

> [T]he mantle of traditions...becomes wrapped around and entangled with a whole new set of people and social processes (Gibson, 1991: 294).

Note

1   Data for this analysis have been distilled from research in the Mackay area of North Queensland (McAllister *et al.*, 1995; Passfield *et al.*, 1996); from a study of small farmers, including cane-growers, in the Bundaberg region of Central Queensland (Cox and Low, 1996); from current research into the Bundaberg industry (Finemore, 1998); and from discussions with industry leaders in both these regions.

References

Alston, M. (1992), *Rural Women*, Centre for Rural Welfare Research, Charles Sturt University: Wagga Wagga.

Anon (1970), *Employment and Technology, No. 7: Men and Machines in Sugar Cane Harvesting*, Dept. of Labour and National Service: Melbourne.

Balazategui, B. (1990), *Gentlemen of the Flashing Blade*, Department of History and Politics, James Cook University: Townsville.

Barlett, P. (1986), 'Profile of Full Time Farm Workers in a Georgia County', *Rural Sociology*, Vol. 50, No. 1, pp. 78-96.

Cox, V. and Low, S. (1996), *Needs Survey Report of the Rural Industry Producers of Bundaberg District 1995-1996*, Agri-business Development Centre, Central Queensland University, Bundaberg.

Davis, J. (1980), 'Capitalist Agricultural Development and the Exploitation of the Propertied Labourer', Buttel, F. and Newby, H. (eds), *The Rural Sociology of the Advanced Societies: Critical Perspectives*, Allanheld: Montclaire, pp. 133-154.

Drummond, I. (1996), 'Sweet and Sour: The Dynamics of Sugar Cane Agriculture', *International Journal of Sociology of Agriculture and Food*, No. 5, pp. 40-65.

Djurfeldt, G. (1996), 'Defining and Operationalizing Family Farming from a Sociological Perspective', *Sociologia Ruralis*, Vol. 31, No. 3. pp. 340-351.

Encel, S. (1970), *Equality and Authority: A Study of Class, Status and Power in Australia*, Cheshire: Melbourne, pp. 293-317.

Evatt Foundation (1991), *Australian South Sea Islanders: A Report on the Current Status of South Sea Islanders in Australia*, Evatt Foundation: Sydney.

Finemore, M. (1998), 'The Social Consequences of Industry Restructuring for Sugarcane Growers and Harvesters Supplying Two Mill Areas in the Bundaberg District', in O'Brien, W., Grasby, D., Brigg, M. and Hungerfod, L. (eds), *Diverse Dialogues: Postgraduate Contributions from Central Queensland University*, Central Queensland University Press: Rockhampton.

Gibson, K. (1991), 'Company Towns and Class Processes: A Study of the Coal Towns of Central Queensland', *Environment and Planning D: Society and Space*, Vol. 9, pp. 285-308.

Gill, H. (1981), 'Land, Labour or Capital: Industrial Relations in the Australasian Primary Sector', *The Journal of Industrial Relations*, Vol. 23, No. 2, pp. 139-162.

Gistitin, C. (1995), *Quite a Colony: South Sea Islanders in Central Queensland, 1867-1993*, AEBIS Publishing: Brisbane.

Goodman, D., Sorj, B. and Wilkinson, K. (1987), *From Farming to Biotechnology*, Basil Blackwell: Oxford.

Gray, I., Lawrence, G. and Dunn, T. (1993), *Coping with Change: Australian Farmers in the 1990s*, Centre for Rural Social Research: Wagga Wagga.

Higgins, V. (1996), *Breaking Down the Divisions? Political Power, Graziers, and Changing Local Government Representation in the Longreach Shire, Queensland*, BA(Hons) Thesis, Department of Social Sciences, Central Queensland University: Rockhampton.

Hungerford, L. (1996), 'Australian Sugar and the Global Economy', in Burch, D., Rickson, R. and Lawrence, G. (eds), *Globalization and Agri-Food Restructuring: Perspectives from the Australasia Region*, Avebury: Aldershot, pp. 127-138.

James, K. (ed.) (1989), *Women in Rural Australia*, University of Queensland Press: St. Lucia.

Jarosz, L. (1996), 'Working in the Global Food System: A Focus for International Comparative Analysis, *Progress in Human Geography*, Vol. 20, No. 1, pp. 41-55.

Lawler, R. (1959), *The Summer of the Seventeenth Doll*, Fontana: London.

Lawrence, G. (1987), *Capitalism and the Countryside*, Pluto Press: London.

Lawrence, G. (1996), 'Contemporary Agri-food Restructuring: Australia and New Zealand', in Burch, D., Rickson, R. and Lawrence, G. (eds), *Globalization and Agri-Food Restructuring: Perspectives from the Australasia Region*, Avebury: Aldershot, pp. 91-103.

Lockie, S. and Vanclay, F. (eds) (1997), *Critical Landcare*, Charles Sturt University: Wagga Wagga.

Lyons, K. (1996), 'Agro-industrialisation and Rural Restructuring: A case study of the Australian Poultry Industry', in Lawrence, G., Lyons, K. and Momtaz, S. (eds), *Social Change in Rural Australia*, Rural Social and Economic Research Centre, Central Queensland University: Rockhampton, pp. 167-177.

May, D. (1994) *Aboriginal Labour and the Cattle Industry: Queensland From White Settlement to the Present*, Cambridge University Press: Melbourne.

McAllister, J., Lawrence, G. and Passfield, R. (1995), *Extended Hours of Cane Harvesting: Findings from Focus Groups, Questionnaire Surveys and Community Based Research in the Mackay Region of Queensland*, Rural Social and Economic Research Centre, Central Queensland University: Rockhampton.

Miller, S. (1996), 'Class, Power and Social Construction: Issues of Theory and Application in Thirty Years of Rural Studies', *Sociologia Ruralis*, Vol. 36, No. 1, pp. 93-116.

Mooney, P. and Majka T. (1995), *Farmers' and Farm Workers' Movements: Social Protest in American Agriculture*, Twayne: New York.

Nankervil, P. (1980), 'Australian Agribusiness: Structure, Ownership and Control', in Crouch, G., Wheelwright, T. and Wiltshire, T. (eds), *Australia and World Capitalism*, Penguin Books: Ringwood, pp. 160-168.

Newby, H. (1974), *The Deferential Worker*, Allen Lane: Harmondsworth.

Parsons, H. (1995), *The Role of the Melbourne Wholesale Fruit and Vegetable Market in Fresh Produce Supply Chains*, Working Paper, No. 35, Department of Geography and Environmental Science, Monash University: Clayton.

Passfield, R., Lawrence, G. and McAllister, J. (1996), 'Not So Sweet: Rural Restructuring and its Community Impact - the Mackay Sugar District', in Lawrence G., Lyons, K. and Momtaz, S. (eds), *Social Change in Rural Australia*, Rural Social and Economic Research Centre, Central Queensland University: Rockhampton, pp. 188-200.

Rickson, R. and Burch, D. (1996), 'Contract Farming in Organizational Agriculture: The Effects upon Farmers and the Environment', in Burch, D., Rickson R. and Lawrence, G. (eds), *Globalization and Agri-Food Restructuring: Perspectives from the Australasia Region*, Avebury: Aldershot, pp. 173-202.

Rowley, C. (1972), *The Destruction of Aboriginal Society*, Penguin Books: Ringwood.

Ruben, A. (1992), *A Study of Seasonal Rural Workers in an Advanced Capitalistic Society. The Fruit Pickers in Victoria, Australia*, unpublished PhD Thesis, La Trobe University: Bundoora.

Smailes, P. (1996), 'Entrenched Farm Indebtedness and the Process of Agrarian Change: A Case Study and its Implications', Burch, D., Rickson, R. and Lawrence, G. (eds), *Globalization and Agri-Food Restructuring: Perspectives from the Australasia Region*, Avebury: Aldershot, pp. 301-322.

Taylor, M. (1997), *Bludgers in Grass Castles: Native Title and the Unpaid Debts of the Pastoral Industry*, Resistance Books: Sydney.

Thorpe, B. (1992), 'Aboriginal Employment and Unemployment: Colonised Labour', Williams, C. with Thorpe, B (eds), *Beyond Industrial Sociology: The Work of Women and Men*, Allen and Unwin: Sydney.

Vanclay, F. and Lawrence, G. (1995), *The Environmental Imperative: Eco-Social Concerns for Australian Agriculture*, Central Queensland University Press: Rockhampton.

van der Ploeg, J. (1992), 'The Reconstitution of Locality: Technology and Labour in Modern Agriculture', in Marsden, T., Lowe, P. and Whatmore, S. (eds), *Labour and Locality: Uneven Development and the Rural Labour Process*, David Fulton Publishers: London, pp. 19-43.

Wells, M. (1984), 'The Resurgence of Sharecropping: Historical Anomaly or Political Strategy', *American Journal of Sociology*, Vol. 90, pp. 1-19.

Wells, M. (1996), *Strawberry Fields: Politics, Class and Work in Californian Agriculture*, Cornell University Press: Ithaca.

Wright, E. (1978), *Class, Crisis and the State*, New Left Books: London.

PART IV
THEORISING KEY CONCEPTS
IN AGRI-FOOD RESEARCH

16 'Feed the man meat': Gendered food and theories of consumption

STEWART LOCKIE AND LYN COLLIE

He's got a man sized hunger, just what food can beat.
So feed the man, feed the man, feed the man meat.
(advertising jingle, Australian Meat and Livestock Corporation, 1985)

Eat more beef you bastards!
(car bumper sticker, Central Queensland, 1997)

Introduction

In the mid-1980s the Australian Meat and Livestock Corporation (AMLC) attempted to redress the accelerating loss of market share as demand for red meats (beef and lamb) declined in the face of increased consumption of chicken and pork.[1] The centrepiece of the campaign was a television advertisement which featured a young boy, clad in the jersey of a well-known football club, running home with his dog through an inner city working class area, for dinner. The boy arrives home in time to join his sister and father (who is reading a newspaper) at the kitchen table, just as his mother removes a large sizzling leg of roast meat from the oven. Through the entire advertisement there runs an upbeat soundtrack and a catchy jingle proclaiming 'feed the man meat'.

According to Reg Bryson, deputy chairman of the Campaign Palace advertising agency, this advert had it all - a catchy jingle, impact, and consumer recognition (Ross-Smith and Walker, 1990). However, it failed miserably. The social context within which food consumption practices were located had changed, rendering those signifiers of patriarchal masculinity, stereotypical gender roles and a 'good hearty meal' (centred around the formula of 'meat and three veg') of declining relevance to those consumers who had reduced their consumption of red meat. Subsequently, the campaign was abandoned and replaced with the promotion of beef and lamb 'short cuts', a series of advertisements, recipe cards and point-of-sale packaging strategies which emphasised how red meat could be integrated into the lifestyles of contemporary consumers, including single and working women and men. Meat was presented as healthy, sophisticated, multicultural, convenient and, perhaps most importantly, as part of a meal rather than a meal in itself (*The Land*, 6 November 1997; see also Lupton, 1996). The 'short cuts' campaign proved popular with consumers, yet meat consumption per capita in Australia has continued to decline. Currently, up to 25 percent of Australian women claim to be strictly or partly vegetarian (Story, 1997), while meat producers stagger from one market or health crisis to the next.

As Dietz *et al.* (1995) point out, increasing rates of vegetarianism - along with more general declines in per capita meat consumption - could be expected to have substantial impacts on established food production, processing and distribution networks. In apparent contradistinction, however, an increasing taste for meat in countries like Japan and Korea (Tokayama and Egaitsu, 1994; *The Land*, 6 November 1997) and the expansion of feedlot production (Lawrence and Vanclay, 1994) promise to expand both international consumption of meat and the production of grains and other feedlot inputs.

No matter how these potentially contradictory processes are worked out, the observation that consumption must be regarded as central to the restructuring of agri-food networks seems so apparent as to be banal. Yet not only do studies of consumption remain few in number within the agri-food restructuring literature, but there has been limited effort in theorising the concept of 'consumption' and its relationship to other practices within agri-food networks. Indeed, as Tovey (1997) points out, an implicit division of labour has become established in relation to the study of food between rural sociologists who study the organisation of agriculture and sociologists of food who focus on eating, diet and culture.[2] In seeking to assess the relevance of consumption studies to our understanding of agri-food restructuring, this chapter argues that the meaning of consumption cannot be taken for granted as simply the obverse of production, nor reduced to the economistic notion of 'demand' (see Fine, 1995). Rather, there is a need to develop methodological and theoretical tools that link the *social practices* associated with food consumption with the many other practices involved in its production, processing, distribution and so on. In beginning this task, this chapter will review the conceptualisation of 'consumption' in both the production-focussed rural sociology and consumption-focussed sociology of food literatures. It will then discuss these in the context of a small pilot study into food consumption and regional identities conducted in Rockhampton - Australia's self-proclaimed 'Beef Capital' - with particular emphasis on the ways in which the gendering of foods shapes consumption and production practices.

Theorising consumption

Agriculture, rural space and consumption

There are four primary ways in which notions of consumption have been integrated into rural sociological discourse. The first of these is Goodman and Redclift's (1991) attempt to locate transformations in the social relations of production and consumption in relation to shifts in the labour process and technology, which saw women move into wage employment accompanied by the introduction of new domestic technologies into the home. Women's domestic labour was thus commoditised and converted into an arena of accumulation, shifting the focus of the home from a site of production to a site of consumption. Goodman and Redclift (1991) suggest that this had far-reaching consequences for food systems as consumption shifted towards processed convenience foods, which promoted product differentiation, the lengthening of production processes and 'value adding'. This, in turn, transformed farmers from the producers of food into the suppliers of inputs to industrial food manufacturing processes, breaking the linkages between food provision, sustainable farm management, and rural society. Goodman and Redclift (1991) acknowledge the phenomenological or symbolic dimensions of food consumption - including their manifestation in alternative food movements - but argue that it is the material basis of food production and consumption that establish new areas of commodity production and, consequently, the values attached to social relations.

The second way in which consumption has been applied has been through the lens of regulation theory. According to regulation theory, relatively stable patterns of production and capital accumulation (regimes of accumulation) are supported by 'institutional forms, societal norms and patterns of strategic conduct' (modes of regulation), that express and regulate conflict in accumulation until crisis points are reached and new arrangements emerge (Jessop, 1988: 149). Marsden *et al.* (1993) argue that the state plays a crucial role in mediating the interdependencies of production and consumption, which has been evident in the shift of policy from support for Fordist forms of mass production and consumption, towards deregulation and privatisation which, in turn, promote flexible specialisation and product differentiation targeted at niche markets. Support for this argument is drawn from Friedmann's (1988; see also Friedmann and McMichael, 1989; Friedmann, 1993) analysis of particular food complexes and her conclusion that Fordist norms of consumption are dependent on nationally and regionally variable class-based consumption habits. The social regulation of consumption is thus held to be vital to the sustainability of any given regime of accumulation. What is rather less clear is why it is state action, in particular, that is of primary importance in maintaining the confluence of production and consumption patterns during stable phases of accumulation.[3]

The third application of a notion of consumption has been via analyses of 'rural' spaces as either sites, or objects, of consumption following the decline of agriculture as *the* determining feature of the social relations of rural areas (Redclift and Whatmore, 1990). Associated by various authors with post-Fordist production (Marsden *et al.*, 1993), postmodernisation (Lawrence, 1995) and mass tourism (Urry, 1995), it is held that rural spaces have been progressively colonised by non-farming activities and thus reconfigured as sites, or objects, of consumption as opposed to sites of agricultural production. This is manifest in the relocation of high technology information industries to rural areas, counter-urbanisation and competition for housing among the geographically and socially mobile, and pressures to maintain the countryside in a visual form that accords with notions of a rustic rural idyll - rather than with contemporary industrial agriculture - rendering it attractive to non-farming residents and visitors. While rural households may become sharply differentiated in terms of their access to the means of consumption - for example, between farming, non-farming and pluriactive households - Redclift and Whatmore (1990) also point out that differentiation may occur within households on the basis of gender inequalities in relation to property rights, labour and surplus value.

The final way in which consumption has been applied within rural sociological discourse has been in the context of 'green consumerism' and alternative food movements. According to Lawrence (1996) 'green consumers' eschew mass produced foods (those understood to be over-processed, over-packaged, containing potentially harmful additives and produced in an environmentally unsustainable manner), and instead demand foods that are clean, nutritious and environmentally 'friendly'. Buttel (1994) argues that it will be the various branches of the environmental movement - incorporating elements of 'green consumerism' - that are likely to pose the most significant challenge to the further industrialisation of agriculture and food processing. This is evident in a range of potentially contradictory processes, including growing demand for organically produced (i.e. without synthetic chemicals or fertilisers) foods (see Lyons, 1997), community supported agriculture, farmers' markets and the development of niche markets in 'health' foods by manufacturers and retailers (Goodman and Redclift, 1991).

It is acknowledged that in sketching out these various perspectives, scant attention has been paid to their more complex arguments and implications. Nevertheless, a couple of points do bear drawing out. Firstly, the only perspective in which the notion of consumption is explicitly theorised is the regulationist reading of Marsden *et al.* (1993), in which consumption is seen to be manipulated by the state in order to provide a stable basis for production. In this respect, consumption is seen as something which is *produced* from the outside, rather than as the simple agglomeration of individual consumer choices. A similar, though implicit, conceptualisation is adopted by Goodman and Redclift (1991) as they argue that the commoditisation of the household and women's labour shifted consumption patterns in the interests of capital. For the two other perspectives, consumption is not determined by production but reflects consumer preferences and choices, although it is acknowledged that such choices are not entirely unfettered and may place consumers in conflict with producers and other consumers. These perspectives thus reflect long-standing dichotomies within sociological theory between conceptualisations of structure and agency, and macro and micro-levels of analysis - replacing these with a dichotomy between the 'production of consumption' and the 'dictatorship of the consumer' (see Miller, 1995). While adopting either of these positions in isolation may reflect understandable attempts to locate *the* locus of control within processes of agri-food restructuring, there is a danger that in doing so multiple spheres of influence may be ignored (see also Lockie, 1998).

Secondly, considerations of consumption by all of these perspectives are overwhelmingly restricted to commodity exchange. This ignores the consumption of non-commoditised goods and services such as domestic labour, the collective consumption of public goods, the meanings associated with any given commodity, and the use, care, maintenance and transformation of commodities following exchange (see Dowding and Dunleavy, 1996; Miller, 1995). This focus on commodity exchange may not be entirely unjustified, but it is clear that the moment of commodity exchange does not by itself exhaust possibilities for the relevance of consumption, or its understanding, to processes of agri-food restructuring. Fine (1995) argues that such limited conceptualisations of consumption reflect the scant interest taken by producers and retailers in the post-exchange life of commodities. However, the meanings which producers and retailers attempt to attach to commodities - their symbolic production (Cook, 1994) - in order to influence the final moment of exchange represent clear attempts to symbolically locate commodities within the social and cultural milieux through which they will be transformed following exchange.

Food, meaning and identity

In contrast to rural sociological discourse, work on the sociology of food has focussed a great deal of attention on the social meanings of food consumption and the practices associated with it, drawing heavily on the work of social anthropologists and historians. According to Campbell (1995: 106), the bulk of anthropologically and historically informed work has tended to divide between a 'fundamentally materialist approach that focuses on food as related to issues of diet or nutrition, on the one hand, and [an approach] which treats food (or 'foodways') as codes or symbolic systems capable of semiotic or structural analysis on the other'. Studies taking the former approach have often done so from within the framework of a medicalised understanding of the body and focused upon issues such as eating disorders (e.g. Bordo, 1997), while studies taking the latter approach have focussed more on the role of food in maintaining group identities and reproducing social structures (e.g. Bourdieu, 1984; Douglas, 1997; Levi-Straus, 1997; Mintz, 1996). In addition, Campbell (1995) identifies another three, more distinctly sociological, approaches. The first of these has used historical analyses to demonstrate how taste and appetite can be shaped by social, political and economic processes (e.g. Mennell, 1985). The second approach has focussed on the structuring of food provision within the household according to age, gender and life-cycle (e.g. Charles and Kerr, 1986), while the third has addressed issues related to consumption and human embodiment (e.g. Bourdieu, 1984; Falk, 1994; Lupton, 1996).

Again, none of these perspectives will be examined in depth here, but it is worth drawing attention to the ways in which the notion of consumption has been conceptualised. The consumption-focussed theorists suggest that the most sociologically significant moment in the consumption of food lies in its appropriation into the human body, rather than its exchange as a commoditised good. Food consumption is seen at one and the same time as both an intensely personal experience through which the senses are stimulated and the elements of food are broken down and incorporated into the human body, and a profoundly social one through which the meanings associated with food incorporate individuals into social groups, ascribe identity and shape subjectivity. According to those authors drawing on the sociology of the body, food does not merely carry meanings in an abstract or purely symbolic sense; those meanings (and the social relationships in which they are embedded) are also physically inscribed upon the human body via food consumption and manifested in characteristics such as size, shape, deportment, strength and so on (Lupton, 1996).

An obvious criticism of this literature is that it either tends to ignore issues related to food production and availability, or to discuss them using a rather simplistic conceptualisation of 'the food system' (see Beardsworth and Keil, 1997). In terms of developing a theory of consumption that is relevant to questions of agri-food restructuring, this is clearly a major deficiency. Nevertheless, I would argue that it is the possibilities that are suggested by this literature for a fundamentally different investigation of the relationships between commoditisation and consumption (i.e. one that does not privilege either production or individual consumer choice), that holds promise for agri-food research. Oppositional food movements may serve as a useful example here. The additives and agricultural chemicals that concern 'green consumers' do not merely threaten some abstract notion of environmental well-being but, rather, threaten to transgress the integrity of consumers' own bodies and senses of self (see Miller, 1995). However, as threats that are difficult to discern with the human senses, people are forced to rely on the claims and counter-claims of risk professionals, 'experts' who calculate the risks these foods present, but who are often seen to represent the interests of those who produce those risks (Beck, 1992). There is no shortage of firms which have moved to capitalise on these fears and uncertainties by attaching signifiers of environmental health and safety to their products. Yet, to reduce the complexities of this phenomenon to a market segment in which capital may pursue new strategies of accumulation would be perverse, and would overlook the extent to which firms have responded to the food 'counter culture' as opposed to developing their own 'counter cuisine' (Belasco, 1993). Importantly, a key focus for many oppositional food movements has been these very commoditised relationships between producers and consumers (Belasco, 1993), a focus that has led to the development of what Miller (1995) argues are fundamentally new forms of consumption. For these movements, reconfiguring the moment of commodity exchange is inseparable from a similar reconfiguration of the moment of ingestion, the cultural milieu within which the ingestion of that food takes place, and its consequences. While relationships between commodity production and exchange, and the meanings and practices associated with ingestion, may not be clearly articulated in the form of a political strategy for all commodities or social groups, both must be incorporated into any coherent conceptualisation of food consumption.

Towards a more holistic theorisation of consumption in agri-food restructuring

The limitation of sociological interest in consumption to the exchange of commodities is certainly not unique to rural sociology (see Bocock, 1993). However, in advocating a cautious attitude to the articulation of narrow definitions of consumption, Miller (1995) argues firstly, that consumption should not be reduced to acts of purchase; secondly, that the meaning of consumption and exchange is historically and culturally variable; but thirdly, that the impersonality and anonymity of highly rationalised institutions and bureaucracies characteristic of modernity, place people in a secondary position in relation to the means of production. Identifying less with the institutions that produce goods and services, people come increasingly to identify themselves as consumers, utilising their consumption of commodities to create specificity and identity (see also Munro, 1996). There can be little doubt that the commodity form is of fundamental importance to contemporary consumption in the advanced economies. There is more than a little resonance though, between Miller's third point and the argument that the ingestion of food is central to the maintenance of social identity and the formation of subjectivity (Lupton, 1996), reinforcing the argument that consumption should not be conceptualised exclusively in relation to the consumption of commodities in the first instance, nor to their exchange in the second.

Asserting that the meaning of consumption is historically and socially variable does not in itself, however, provide us with many clues as to how to go about understanding it even within particular social formations. Nor does the argument also suggested above, that neither producers nor consumers should be accorded *a priori* a determinant role in shaping the relations of consumption. A notable attempt to develop a methodology that goes some way to resolving these dilemmas has been provided by Fine (1995; see also Fine and Leopold, 1993; Fine *et al.*, 1996), who argues that a shift is necessary from 'horizontal' to 'vertical' analyses. Horizontal analyses, he suggests, examine single commodities from narrow disciplinary perspectives and then generalise their results to the nature of consumption in general. Vertical analyses, by contrast, examine particular commodities - or groups of commodities, 'in the context of the chain of horizontal factors that give rise to [them] - production, distribution, retailing, consumption and the material culture surrounding [them]' (Fine, 1995: 142). While these separate factors are common to many commodities, the unique ways in which they interact contribute to the development of equally unique 'systems of provision' for each commodity or commodity group which cannot be generalised to other commodities. At first glance this appears to have much in common with 'commodity systems analysis' as developed by Friedland (1984), with its focus on production practices, grower organisation, labour, science production and application, marketing and distribution and the role of the state. Key additional factors considered by Fine (1995) however, include the material culture surrounding commodities and the role and agency of consumers within systems of provision, although I would suggest that the consideration of these factors in the studies of sugar and meat by Fine *et al.* (1996) are weak due to a preference for macro-level statistical methods.

The fact that Fine (1995) does not acknowledge the consumption of non-commoditised goods and services may limit the utility of this approach, but there are at least two reasons why this should not be considered detrimental. The first of these is the likelihood that many non-commoditised practices involving food may be considered as part of the material culture that surrounds its commoditised forms and may be studied on that basis. The second and related point is the contemporary ubiquity of the commodity form and commodity exchange. This ensures that not only do people interact constantly with commodities, but that the meanings surrounding even non-commoditised practices often take commodities or their 'systems of provision' as something of a reference point, even if only in opposition to them.[4] Nevertheless, by privileging analysis of commodity chains, Fine's articulation of 'vertical analysis' is limited in its ability to incorporate socio-cultural aspects of food consumption that are not structured around single commodities (including vegetarianism, distinct ethnic practices, the organics movement and so forth), but that interact nevertheless with commoditised systems of provision in potentially important ways. These aspects of food consumption are integral to its material culture, production and consumption, but are indivisible to discrete elements that can be apportioned across individual commodities prior to detailed analysis. Further, the spatial and temporal variability of socio-cultural aspects of food consumption is not likely to correspond very neatly with the boundaries of globally distanciated networks of commodity exchange. Clearly, the need for detailed horizontal analysis still exists in order to draw out the implications of *particular food cultures* for, perhaps, more than one particular commodity or commodity group. Again, this does not mean that 'vertical analysis' should be abandoned, but that an approach more open to detailed ethnographic inquiry would be better able to deal with the whole material culture surrounding food production and consumption. Rather than limiting such inquiry to exclusively micro-level analysis, the application of 'multi-focal' ethnographies' (Marcus 1992; see also Jackson and Thrift, 1995), in combination with political economic analysis, would allow greater integration of the material and symbolic elements of food consumption, and greater appreciation of the relationships between commoditised and non-commoditised forms of consumption.

Meaning, masculinity and meat

> There are currently 1.28 billion cattle populating the earth. They take up nearly 24 percent of the landmass of the planet and consume enough grain to feed hundreds of million of people. Their combined weight exceeds that of the human population on earth (Rifkin, 1992: 1).

Cattle - the source of the most prestigious of all commonly consumed meats - are clearly of enormous social, economic and ecological importance. Sheep are not insignificant either. Together, the flesh of these animals carries such high status that it is a metonym for the idea of food itself, and central to most peoples idea of a 'proper meal' (Lupton, 1996). Yet, the most ubiquitous symbolic feature of red meat consumption is not its environmental, nutritional or economic consequences, but rather its association with masculinity. Drawing on a number of studies that have examined cultural aspects of meat consumption, it is possible to argue that for western societies, in a very general sense, meat is symbolic of virility, strength, aggressiveness, power, status, lustfulness, energy and health, while vegetables are symbolic of 'purity, passiveness, cleanliness, femininity, weakness and idealism' (Lupton, 1996: 28; see also Beardsworth and Keil, 1997; Fiddes, 1991). Meat is also, however, the source of profound ambivalence amongst its consumers. At the same time that meat consumption promises health and vitality, it threatens with contamination and decay - from bacterial infection, chemical residues and *bovine spongiform encephalopathy* (or 'mad cow' disease) to cardiovascular disease and cancer (Lupton, 1996). While the bloody and violent origins of meat arouse unease among many, they are also a potent symbol of human domination over raw nature and the ability of humans to control and appropriate their environment

(Fiddes, 1991). Indeed, were the sourcing of meat less brutal, Fiddes (1991) argues, it would lose much of its symbolic force. This representation of masculinity, domination and control through the killing and dismemberment of animals is also symbolic, Adams (1990) argues, of women's objectification and subjugation by men. At the same time though, the symbolic promise of power and virility in the blood that drips from red meat may itself prove too strong, perhaps transgressing taboos against consuming raw meat or the flesh of certain animals (Beardsworth and Keil, 1997).

According to writers such as Lupton (1996), growing numbers of consumers are somewhat repulsed by the signification by red meat of masculine power and the domination of nature. For these consumers, the ambivalence they feel towards the consumption of red meat is increasingly resolved in favour of either its replacement with symbolically less potent meats such as chicken and fish, or a general reduction or rejection of animal flesh within their diets. Growing unease among consumers about the violent origins of meat has also been held responsible for changes in retailing practices which avoid confronting consumers with recognisable animal carcasses and package meat as meals rather than body parts. It is, of course, possible that violent and bloody imagery still appeals to a great many red meat consumers, but these are not the people that the meat industry needs to influence if they are to arrest the decline in consumption. However, contradicting these explanations for declining red meat consumption, the Australian Meat and Livestock Corporation contends that the main factor lies in the wider array of alternatives that are now available to consumers (Story, 1997). According to this suggestion, people are not concerned about ingesting red meat itself, but merely wish to explore a wider range of gustatory pleasures. It suggests that, reflecting growing cosmopolitanism, people are simply more inclined to experiment, to ask themselves, why not try tofu or tempeh or, for that matter, emu or kangaroo?

Meat consumption in Rockhampton, the 'Beef Capital' of Australia

It is stating the obvious to suggest that the reasons behind changing consumption practices in relation to red meat are of fundamental importance to further restructuring of both the meat industry and the communities that are dependent upon it. The city of Rockhampton and its rural hinterland are prime examples. In 1994, the Central Queensland region had a human population of 308,615 compared to a bovine population of some 3.2 million. Rockhampton, a service centre of 62,741 people, was the site for two of the regions four export abattoirs.

It was also, however, the most slowly growing centre in Central Queensland, reflecting its economic dependence on the struggling beef industry. While this paper cannot make a definitive judgement on the conflicting explanations offered above for the decline in red meat consumption, the research presented in this section explores these issues in the context of a study of meanings associated with food consumption and locality in Rockhampton. This research utilised a series of four focus groups (two exclusively female and two exclusively male), and five in-depth interviews with long-term residents of Rockhampton, to explore changes in food consumption practice over the lifetime of participants (all data from Australian Bureau of Statistics, 1996; Vercoe, 1996).

The 'normal' diet

In general terms, the results of this research were consistent with those reported by Lupton (1996), Fiddes (1991) and Beardsworth and Keil (1997). Participants reported growing up on what they saw as simple and 'normal' diets which included large quantities of both fresh and preserved meat. Bread and dripping[5] was a common snack or lunch, while vegetables, fruit, eggs and milk were often produced by the family for their own consumption or bartering. Doug, 66, the one vegetarian involved in the study stated:

> I think in those days people used to laugh at rabbits, at lettuce eaters, but now there's a lot of people…turning to vegetarianism that have no interest in religion whatsoever. They're looking at it for health. Vegetarianism's sorta coming into it's own now. But in those days yeah, I think people really thought you were a bit strange. But most people were too polite to point it out …

Despite Doug's experience of a growing acceptance of vegetarians, and the commoditisation of the aforementioned subsistence activities, the centrality of meat to the diets of other participants remained largely unchanged. What did change however, were the social practices and meanings surrounding the moment of ingestion and the quantity and type of meat consumed. A key feature of contemporary consumption according to many participants, was the variety of foods available and the opportunities they afforded for exploration and experimentation, but again this focussed on different ways of preparing meat as opposed to discarding it.

Gendering food practices: Chefs and hunters

Also consistent with other studies, the practices and meanings surrounding meat consumption were clearly gendered. This was perhaps most evident in relation to food preparation within the home, even despite men's self-reported willingness to cook. Women discussed the ways in which they devoted their creativity when preparing food to keeping their husbands and children happy, a practice which often constrained their 'adventurousness' and saw them prepare more red meat than they may have liked to consume themselves. Conversely, when men discussed food preparation it was primarily in the context of entertaining, and preparing either elaborate banquets or backyard barbeques. Women constructed themselves as mothers and wives while men constructed themselves as chefs and entertainers; the former roles were firmly located within the private sphere of domestic labour and servitude, and the latter in the public spheres of work and sociability.

Ambivalencies around meat consumption also took on gendered dimensions, although there was little obvious evidence of the sort of revulsion or unease reported by other studies about the origins of meat. Many participants in this research had had childhood experiences of the slaughtering of animals to eat at home and, for men in particular, fishing and hunting were popular recreational pursuits. Indeed, fishing and hunting were spoken about with such great passion amid the appreciative murmurs of other focus group members, that it seemed there was a general consensus that these were regarded as among life's greatest pleasures. Rifkin (1992) suggests that the modern cattle complex serves as a visible reminder of a distant past characterised by the violence of the hunt, colonisation and greed, but for participants in this research the hunt and the consumption of its quarry remained a very contemporary experience. Women also reported participating in fishing and enjoying the spoils of their catch, but with less enthusiasm and less interest from other group members, suggesting that, while they were not excluded from practices associated with killing, it remained a particularly masculinised practice. Without more detailed ethnographic research it is difficult to do more than speculate as to how different practices associated with killing contributed to the ways in which masculinity was constructed, but it does seem likely that for these men, participation in these practices helped to reinforce a peculiarly masculine sense of self. Perhaps surprisingly then, those men who had spent their working lives on the killing floor of one of the local abattoirs thought slaughtering livestock such a gruesome practice (especially pigs because of their human-like screams) that they became *immune* to it through exposure, as opposed to something that they revelled in and enjoyed in the same way they did fishing. Perhaps some unease about the violence of bloodletting is evident here, which was not confronted to the same extent by killing fish as cattle, but it is also likely that the ability to stomach the more gruesome aspects of such practices is also constructed as a particularly masculine attribute.

Gendered ingestion and the body

On the other hand, revulsion was evident in relation to the consumption of offal, with many participants describing being forced to eat it as children (see also Lupton, 1996) and declaring their horror at the human-like attributes of some of these body parts. One participant stated that they would not eat bullocks tongues 'because tongues speak'. Another described liver as a food with 'thinking power', but one of their fellow group members interjected with the suggestion that,

> it's thinking 'why are you eating me?'

These exchanges are suggestive of an association made by participants between the consumption of higher order mammals and the cannibalism invoked by the consumption of body parts recognisably similar to those of humans. According to Fiddes (1991), it is this evolutionary closeness of higher order mammals to humans that is fundamental to the powerful meanings of domination over nature attached to meat consumption, but there was little evidence here of such an association among those participants who did enjoy consuming offal.

By far the greatest source of ambivalence about meat consumption, however, and the one most clearly linked to changing consumption practices, was its potential health effects. For some people this related to issues of quality - such as the presence of chemical residues - but mostly it related to the fat content of meat and its possible relationships with cardiovascular disease and body weight. Further, despite the close physiological links between cardiovascular disease and body weight, there was a clear difference in the ways that the male and female bodies were constructed in relation to these so-called 'pathologies'. Both women and men reported being advised by their doctors to modify their diets in the interest of health, but it was only women's bodies that were discussed in terms of their physical attractiveness and the practice of losing weight for its own sake. The men in the study, on the other hand, were aware that they were predominantly overweight, but related this solely to the health problems that might result from being overweight. Indeed, some men were concerned about the possibility of losing too much weight in later life in case they needed reserves of energy to see them through illnesses. In response to these concerns, most participants did claim to have drastically reduced their consumption of fat, but at the same time resisted the medicalised constructions of their bodies that had 'forced' them to do so. This was most immediately evident in the rejection by many of the idea that diet contributed much to body shape at all, suggesting instead that it was an intrinsic genetic feature of each individual.

More important though, was the continuing construction of 'normal' food as the 'simple' meat-based diets of participants' childhoods, in which no attempt was made to reduce fat consumption through, for example, the trimming of visible fat from meat or the removal of skin from poultry. For most, it was believed necessary to at least occasionally, if not frequently, indulge in these diets in order to continue enjoying food and life. To completely discard a diet rich in animal fat - that is, to accept the medicalised construction of their own bodies - was for a few, something they believed that they 'just had to do', implying occasionally that the masculine course of action was to simply accept it and get on with life.

For some others - particularly among the women - changing diets was a constant struggle against the appeal of 'naughty' foods and the contradictory meanings associated with them, while for others - particularly among the men - cutting down on fat was something that they clearly stated they would only do selectively in order to maintain control over their bodies. For these men it was not a fat-rich diet that was demonised, but a sedentary lifestyle induced by either retirement or contemporary work practices. The masculine body was regarded as a body capable of, and characterised by, hard physical work that was sustained and unaffected by a 'normal' fat-rich diet. Even though almost all participants had long abandoned such masculinised hard work, they appeared to be a long way from abandoning the diet they saw as 'naturally' accompanying it.

Conclusion

Reflecting its rural sociological roots, agri-food research has been dominated by political economic analysis of food production and distribution. In attempting to conceptualise the seemingly obvious importance of consumption for these concerns, this chapter has argued that 'consumption' should not be regarded as simply the obverse of production, nor its analysis restricted to the moment of exchange of commodities. This is particularly evident in relation to food, for which the moment of ingestion is of crucial importance in locating people socially and culturally, and which is itself located within complex networks of social practices that together contribute to the whole - albeit fluid and contested - material culture of food. Drawing on Fine's (1995) notion of 'vertical' analysis based on the entire 'system of provision' of a commodity or commodity group, it is suggested that greater integration of commodity systems analysis with 'multi-focal' ethnographies incorporating key moments and sites in the production and consumption of foods, will go some way to the development of better understandings of consumption that do not privilege the agency and power of either producers or consumers. Nevertheless, it is also suggested that detailed 'horizontal' analysis is warranted to explore important food cultures that do not neatly correspond with distinct commodity groups, yet still have potentially important consequences for those commodities.

The exploration of meat consumption presented in this paper is necessarily an incomplete one. Nevertheless, it does serve to illustrate that trend towards reduced per capita red meat consumption in Australia is neither uniform across social groups, nor based on the same sorts of reasons. While other ethnographic studies of red meat consumption have drawn attention to a high degree of ambivalence concerning the bloody origins of meat, this particular group of participants - drawn from a community with close ties to the beef industry and with ready proximity to recreational opportunities including hunting and fishing - were rather more accepting of the killing of animals. Concerns with health were clearly of greater importance to participants, but it was also evident that these concerns were mediated through a number of competing constructions of the human body. These were manifest both in differences between the concerns of women and men participating in the study, and in a highly ambivalent attitude towards medicalised constructions of the body that threatened notions of self-control, normality and masculinity. These aspects of the material culture surrounding the consumption of red meat have clear implications for the production, processing, distribution and retailing of red meat. Yet it is also obvious that the people represented here constitute only a small fraction of the potential consumers even of Australian beef, suggesting that a more holistic understanding of contemporary meat consumption would require both detailed political economic analysis of where and how meat is produced and sold, and the extension of ethnographic inquiry into those 'markets' targeted for increased consumption. For example, it is more or less taken for granted by Australian primary industries that worldwide demand for red meat will increase dramatically in line with the 'Westernisation' or 'Americanisation' of diets in Asia. But few attempts have been made to understand how the material and cultural products of the west have been interpreted, utilised and transformed by these peoples in the context of their own cultural milieux. Such understandings would not only be interesting for their own sake, but would provide crucial clues as to the likely future of food producers and the likely success of various attempts to produce consumption through the attachment of meaning to commodities.

Notes

1   The authors are grateful for financial assistance provided by the Faculty of Arts, Central Queensland University.
2   This lack of integration is illustrated by a recent text by Beardsworth and Keil (1997), in which the rather startling claim is made that 'food system' research has concentrated overwhelmingly on consumption, leaving only a fragmented literature on production.
3   Marsden *et al.* (1993) do illustrate their argument with a more detailed case study of changes in the production and consumption of public housing in Britain in the 1980s. However, while illustrative of the theoretical point they are trying to make, the implicit argument that similar processes are occurring in relation to food production and consumption is not substantiated. Further, it could be argued in contradistinction, that few states have the level of influence over food that the British government had over its own housing programs. It is necessary, therefore, to analyse state action in the context of a much broader range of social processes and actors, some of which may be of far greater significance than the state.
4   Organic and low-input farming provide useful examples in that while based on principles such as the recycling of nutrients and energy, and attracting many adherents who reject the commoditisation and globalisation of systems of food provision, they are actually defined in relation to the commoditised inputs that they do not use (i.e. synthetic chemicals and fertilisers).
5   The solidified fat, or 'pan juices', left over after roasting meat.

References

Adams, C. (1990), *The Sexual Politics of Meat: A Feminist-Vegetarian Critical Theory*, Continuum: New York.
Australian Bureau of Statistics (ABS), (1996), *Queensland Year Book*, Australian Bureau of Statistics: Brisbane.
Beardsworth, A. and Keil, T. (1997), *Sociology on the Menu: An Invitation to the Study of Food and Society*, Routledge: London.
Beck, U. (1992), *Risk Society: Towards a New Modernity*, Sage: London.
Belasco, W. (1993), *Appetite for Change: How the Counterculture Took on the Food Industry*, Updated Edition, Cornell University Press: Ithaca.
Bocock, R. (1993), *Consumption*, Routledge: London.

Bordo, S. (1997), 'Anorexia Nervosa', in Counihan, C. and van Esterik, P. (eds), *Food and Culture: A Reader*, Routledge: London.

Bourdieu, P. (1984), *Distinction: A Social Critique of the Judgement of Taste*, Routledge and Kegan Paul: London.

Buttel, F. (1994), 'Agricultural Change, Rural Society and the State in the Late Twentieth Century: Some Theoretical Observations', in Symes, D. and Jansen, A. (eds), *Agricultural Restructuring and Rural Change in Europe*, Wageningen Agricultural University: Wageningen.

Campbell, C. (1995), 'The Sociology of Consumption', in Miller, D. (ed.), *Acknowledging Consumption: A Review of New Studies*, Routledge: London.

Charles, N. and Kerr, M. (1986), 'Food for Feminist Thought', *Sociological Review*, Vol. 34, No. 3, pp. 537-572.

Cook, I. (1994), 'New Fruits and Vanity: Symbolic Production in the Global Food Economy', in Bonanno, A., Busch, L., Friedland, W., Gouveia, L. and Mingione, E. (eds), *From Columbus to ConAgra: The Globalisation of Agriculture and Food*, University Press of Kansas: Lawrence, pp. 232-248.

Dietz, T., Frisch, A., Kalof, L., Stern, P. and Guagnano, G. (1995), 'Values and Vegetarianism: An Exploratory Analysis', *Rural Sociology*, Vol. 60, No. 3, pp. 533-542.

Douglas, M. (1997), 'Deciphering a Meal', in Counihan, C. and van Esterik, P. (eds), *Food and Culture: A Reader*, Routledge: London.

Dowding, K. and Dunleavy, P. (1996), 'Production, Disbursement and Consumption: The Modes and Modalities of Goods and Services', in Edgell, S., Hetherington, K. and Warde, A. (eds), *Consumption Matters: The Production and Experience of Consumption*, Blackwell: Oxford.

Falk, P. (1994), *The Consuming Body*, Sage: London.

Fiddes, N. (1991), *Meat: A Natural Symbol*, Routledge: London.

Fine, B. (1995), 'From Political Economy to Consumption', in Miller, D. (ed.), *Acknowledging Consumption: A Review of New Studies*, Routledge: London.

Fine, B., Heasman, M. and Wright, J. (1996), *Consumption in the Age of Affluence: The World of Food*, Routledge: London.

Fine, B. and Leopold, E. (1993), *The World of Consumption*, Routledge: London.

Friedland, W. (1984), 'Commodity Systems Analysis: An Approach to the Sociology of Agriculture', in Schwarzweller, H. (ed.), *Research in Rural Sociology and Development*, JAI Press: Greenwich, Vol. 1, pp. 221-236.

Friedmann, H. (1988), 'Family Wheat Farms and Third World Diets: A Paradoxical Relationship Between Waged and Unwaged Labour', in Collins, J. and Giminez, M. (eds), *Work Without Wages: Comparative Studies of Housework and Petty Commodity Production*, State University of New York Press: Binghamton, NY, pp. 193-213.

Friedmann, H. (1993), 'The Political Economy of Food: A Global Crisis', *New Left Review*, No. 197, pp. 29-57.

Friedmann, H. and McMichael, P. (1989), 'Agriculture and the State System: The Rise and Decline of National Agricultures, 1870 to the Present', *Sociologia Ruralis*, Vol. 29, No. 2, pp. 93-117.

Goodman, D. and Redclift, M. (1991), *Refashioning Nature: Food, Ecology and Culture*, Routledge: London.

Jackson, P. and Thrift, N. (1995), 'Geographies of Consumption', in Miller, D. (ed.), *Acknowledging Consumption: A Review of New Studies*, Routledge: London.

Lawrence, G. (1995), *Futures for Rural Australia: From Agricultural Productivism to Community Sustainability*, Rural Social and Economic Research Centre, Central Queensland University: Rockhampton.

Lawrence, G. (1996), 'Rural Australia: Insights and Issues From Contemporary Political Economy', in Lawrence, G., Lyons, K. and Momtaz, S. (eds), *Social Change in Rural Australia*, Rural Social and Economic Research Centre, Central Queensland University: Rockhampton.

Lawrence, G. and Vanclay, F. (1994), 'Agricultural Change in the Semiperiphery: The Murray-Darling Basin, Australia', in McMichael, P. (ed.), *The Global Restructuring of Agro-Food Systems*, Cornell University Press: Ithaca, pp. 76-103.

Levi-Straus, C. (1997), 'The Culinary Triangle', in Counihan, C. and van Esterik, P. (eds), *Food and Culture: A Reader*, Routledge: London.

Lockie, S. (1998), 'Landcare and the State: 'Action at a Distance' in a Globalised World Economy', in Burch, D., Lawrence, G., Rickson, R. and Goss, J. (eds), *Australasian Food and Farming in a Globalised Economy: Recent Developments and Future Prospects*, Monash Publications in Geography No. 50, Monash University: Melbourne.

Lupton, D. (1996), *Food, the Body and the Self*, Sage: London.

Lyons, K. (1997), 'What Shade for the Greenwash', Paper presented at *International Conference on Environmental Justice*, Melbourne, July.

Marcus, G. (1992), 'Past, Present and Emergent Identities: Requirements for Ethnographies of Late Twentieth Century Modernity Worldwide', in Lash, S. and Friedman, J. (eds), *Modernity and Identity*, Basil Blackwell: Oxford, pp. 309-330.

Marsden, T., Murdoch, J., Lowe, P., Munton, R. and Flynn, A. (1993), *Constructing the Countryside*, Westview Press: Boulder.

Mennell, S. (1985), *All Manners of Food: Eating and Taste in England and France from the Middle Ages to the Present*, Blackwell: Oxford.

Miller, D. (1995), 'Consumption as the Vanguard of History: A Polemic by Way of an Introduction', in Miller, D. (ed.), *Acknowledging Consumption: A Review of New Studies*, Routledge: London.

Mintz, S. (1996), *Tasting Food, Tasting Freedom: Excursions in Eating, Culture and the Past*, Beacon Press: Boston.

Munro, R. (1996), 'The Consumption View of Self: Extension, Exchange and Identity', in Edgell, S., Hetherington, K. and Warde, A. (eds), *Consumption Matters: The Production and Experience of Consumption*, Blackwell: Oxford.

Redclift, N. and Whatmore, S. (1990), 'Household, Consumption and Livelihood: Ideologies and Issues in Rural Research', in Marsden, T., Lowe, P. and Whatmore, S. (eds), *Rural Restructuring: Global Processes and Their Responses*, David Fulton: London, pp. 182-197.

Rifkin, J. (1992), *Beyond Beef: The Rise and Fall of the Cattle Culture*, Dutton: New York.

Ross-Smith, A. and Walker, G. (1990), *Women and Advertising: Resource Package*, Office of the Status of Women, Department of the Prime Minister and Cabinet: Canberra.

Story, D. (1997), 'Research Finds 25 Percent of Women Vegetarian', *Queensland Country Life*, 9 May, p. 11.

The Land, 6 November 1997.

Tokayama, H. and Egaitsu, F. (1994), 'Major Categories of Changes in Food Consumption Patterns in Japan 1963-91', *Oxford Agrarian Studies*, Vol. 22, No. 2, pp. 191-202.

Tovey, H. (1997), 'Food, Environmentalism and Rural Sociology: On the Organic Farming Movement in Ireland', *Sociologia Ruralis*, Vol. 37, No. 1, pp. 21-37.

Urry, J. (1995), *Consuming Places*, Routledge: London.

Vercoe, J. (1996), 'The Central Queensland Beef Industry: Swot's it all About?', in Cryle, D., Griffin, G. and Stehlik, D. (eds) *Futures for Central Queensland*, Rural Social and Economic Research Centre, Central Queensland University: Rockhampton.

17 Contract farming in the New Zealand wine industry: An example of real subsumption

JASON MABBETT AND IAN CARTER

Introduction

A cynic might say a contract is an agreement which is binding on the weaker partner only. Increasingly, however, contracts associated with agricultural production represent 'a crucible in which innovative forms of social integration and agrarian corporatism link growers to state and private capital' (Watts, 1992: 90). In particular, contract farming incorporates and subordinates family farms to the interests of capital. The aim of this chapter is to show how contract grape growing in the New Zealand wine industry can be construed as an example of the real subsumption of labour under capital. To do this, we will first define contract farming. Second, we will reiterate Marx's distinction between real and formal subsumption under capital. Third, we will argue that contract farming is an example of real subsumption. Finally, we will illustrate this contention with reference to contract grape growing in the New Zealand wine industry in general, and with particular reference to the case brought by the New Zealand Grape Growers Council to the Commerce Commission in 1991.

Contract Farming

As Watts (1992: 91) notes, contract farming

> entails relations between growers and agro-capitals which substitute for open-market exchanges by linking 'independent family farmers' of widely variant assets with a central processing, export or purchasing unit which regulates price, production practices and credit arranged in advance under contract.

275

Contract farming brings a number of benefits for capital. These primarily derive from capital's attempt to address the biological basis of agricultural production, and the obstacles that this poses to the development of a fully capitalist agriculture (Goodman *et al.*, 1987; Mann, 1990). First, capital seeks to reduce the production risks inherent in agriculture by utilising direct producers,

> while capital saturates and controls the production process through appropriation (machines, seeds, biotechnologies, credit)...[Second] the contract is a legal and institutional means by which appropriationism ramifies throughout global systems of production...the contract that stipulates specific technologies and inputs is an important juridical and institutional means by which appropriation advances onto the terrain of agro-food production (Little and Watts, 1994: 250).

Little and Watts (1994) note that the extent to which appropriation is facilitated by contract farming varies by sector, commodity and region. In particular 'the proliferation of contracting is characteristically associated with the obligatory use of chemical, biotechnological and mechanical inputs and with the industrial processing of contracted outputs' (Little and Watts, 1994: 250).

We dispute the assertions that the relationship between the contract farmer and the contractor are potentially symbiotic (Workman, 1993: 29). Contract farmers are formally integrated with, and subordinated to, circuits of capital. Consequently, rather than being an independent entity, the contract farmer is:

> little more than a propertied labourer, a hired hand on his or her land. Contract farming is less a means to underwrite the preservation of small-scale commodity production, than rather a vehicle to introduce new distinctive work routines, new on-farm technologies and labour processes, to promote a further concentration and centralisation of capital in agro-food systems, and not least to deepen the process of appropriation by which rural farm production processes (farm inputs and services) are converted into industrial capitals and subsequently reincorporated into agriculture. Contract farming marks a critical transformation and recomposition of the family farm sector as capital saturates the entire agro-industrial complex, converting growers into 'self-employed proletariat' without directly taking hold of the point of production (Watts, 1992: 91).

Watts (1992) takes contract farming to be a mechanism which serves to deepen the appropriation of rural production processes. We accept this, but would argue that it also represents an example of real subsumption. This view is contentious, and needs some justification.

The distinction between formal and real subsumption

In what was originally planned to be part seven of volume I of *Capital*, entitled *Results of the Immediate Process of Production*, Marx points to:

> the synthesis of the capitalist mode of production as production of surplus-value and production of commodities produced by capital, and to the interconnected problem of the origin and content of the increased productivity of labour without which no increase in surplus-value production would be possible in the long term (Mandel in Marx, 1976: 944).

To understand this problem Marx makes a distinction between the formal and the real subsumption of labour under capital. According to Stoler (1987), this distinction refers to the qualitatively distinct ways in which capital subordinates labour. It is important to note that subsumption can characterise the way in which capitalism transforms the labour process, incorporates a labouring population or subordinates whole societies to the logic of its reproduction. The distinction therefore 'has been understood as a difference in the degree to which the social relations of production are restructured under capital' (Stoler, 1987: 544).

The formal subsumption of labour under capital

Formal subsumption may be defined as the takeover by capital of a mode of labour established prior to the emergence of capitalist relations (Marx, 1976). For example, according to Marx (1976: 1021), handicraft (the equivalent of a small, independent peasant economy) may be subsumed under capital. This does not presuppose radical changes to the established labour process. Any change is gradual. Working days may be extended, work may become more intensive or orderly, but these changes do not affect the character of the labour process. The capitalist mode of production has a number of mechanisms to extract surplus value, but the pre-existing mode of labour can only increase absolute surplus value by lengthening the working day. 'In the *formal* subsumption of labour under capital this is the *sole* manner of producing surplus-value' (Marx, 1976: 1021). Since labour power is not a commodity, its value cannot be reduced by advances in production, thereby increasing surplus value. Given this fundamental constraint, compulsion is applied. Thus, the essential features of formal subsumption which arise in the subordination under capital are:

> (1) an *economic* relationship of supremacy and subordination, since the consumption of labour-power by the capitalist is naturally directed and supervised by him; [and] (2) labour becomes far more continuous and intensive, and the conditions of labour are employed far more economically (Marx, 1976: 1026).

The real subsumption of labour under capital

Constructed upon the foundation of formal subsumption, a new mode of production - capitalist production (production of commodities including labour power) - transforms the labour process and its actual condition irrevocably (Marx, 1976). Capitalist production establishes itself as a mode of production *sui generis*.Only when this happens does the real subsumption of labour under capital take place. Real subsumption makes possible the extraction of relative surplus value, realised through increased labour productivity. Hence 'it by no means suffices for capital to take over the labour-process in the form under which it has been historically handed down' (Marx, 1976: 298). For Goodman and Redclift (1985: 239):

> This transition, which in industry typically would be associated with a more complex division of labour and large-scale production, involves the real subsumption of labour since the new labour process is beyond the capacity of workers operating as self-employed producers.

The distinction between formal and real subsumption can be used to show how the family farm is structurally integrated within the capitalist mode of production. In particular it 'lays important foundations for the understanding of the impact of capital on agriculture and of family farm production as a particular yet integral part of the capitalist mode of production, not its negation' (Alavi and Shanin, 1988: xxxiii). However, the question still remains; can Marx's distinction be extended to include contract farming?

Mann (1990) argues that contract farming does not comprise capitalist exploitation (the extraction of surplus value at the point of production), but represents another form of exploitation. This is a consequence of various distinctions between contract farmers and wage labourers: most notably that contract farmers own their farms, and use family labour in their production processes. Second, in most cases contract farmers maintain formal ownership of the commodities they produce. This ownership, Mann (1990) argues, is the basis for the contract. The entrepreneur is not purchasing the farmers' labour power, but the commodities the farmer produces with it. Consequently, these commodities constitute what Marx referred to as circulating capital. Thus, '[s]ince labour is not hired, there is no basis for capitalist exploitation proper and even unequal exchange is dubitable because these capitalists must pay the prevailing market price for their contracted commodities' (Mann, 1990: 144).

Against this position, Davis (1980) argues that the family farm in general, and contract farming specifically, should be 'understood not as survivals of a pre-capitalist past, but as organizational components of an unfolding future' (Davis, 1980:135). Non-farm capitalist enterprises contractually establish relations of production with 'independent' agricultural producers. This allows the capitalist to have both direct and indirect control over the production process. Direct control is exerted through the removal of the farmer's exclusive control over decisions and resources. This includes decisions concerning inputs, production and the variety of crop to be grown. The purpose of this direct control is 'increased productivity, as the capitalist firm attempts to ensure that the contract farmer's potential for productive labor is realized as completely and intensely as possible' (Davis, 1980: 143). Indirect control ensures that as a result of the contract (whereby the contract farmer is paid a set price for the agricultural commodities s/he produces), the family farmer will proactively seek to increase productivity and to cheapen the product's unit cost. Consequently,

> the capitalist contractor, in dealing with his time-wage employees, faces serious constraints in his efforts to increase the production of surplus value. He faces few such constraints in dealing with his piece-wage employees - 'independent' family farmers producing under contract, a propertied labor force that is non-unionized, self directed, and willing to work without the guarantees of minimum wage, job security, insurance, and other benefits commonly demanded by time-wage employees (Davis, 1980: 143).

Not content with exerting control over the productive process, capital also controls the off-farm exchange process. Non-farm capitalists determine the market price for products which farmers produce.

The contract thus functions as a mechanism for the extraction of surplus value. This relegates the contract farmer to the position of a disguised wage labourer. The effectiveness of the contract as a means of extracting surplus value, Davis (1980: 144) argues, rests upon two conditions. The first is the degree to which the contractor is unilaterally able to determine the contract price. The second is the degree to which farmers' access to product markets is limited or restricted. As a result:

> The essence, then, of contract farming is control. Through contracting the capitalist firm is able to control both the on-farm production process, maximizing productivity, and the off-farm exchange process, maximizing exploitation. Contract farming is a self-reinforcing capitalist labor process, promoted by the non-farm capitalist firm to assure the production and appropriation of surplus value (Davis, 1980: 144).

Mann (1990) argues that Davis (1980) presents an account which, while purporting to be grounded in Marxist theory, in fact leans on unequal exchange theories grounded in the writings of Ricardo and Weber. She argues that Davis ignores Marx's distinction between capitalist exploitation and other forms of exploitation. This contention, we would argue, is spurious. It may be rebutted at a number of levels. Our first reservation is theoretical. One should not abstract from Marx a disembodied general 'theory' of society or history, which can be mechanically 'applied'. Rather,

> in order to develop the laws of bourgeois economy...it is not necessary to write the *real history of the relations of production*. But the correct observation and deduction of these laws, as having themselves become in history, always leads to primary equations - like the empirical numbers e.g. in natural science - which point towards a past lying behind this system. These indications, together with a correct grasp of the present, then also offer a key to the understanding of the past - a work in its own right which, it is to be hoped, we shall be able to undertake as well. This correct view likewise leads at the same time to the points at which the suspension of the present form of production relations gives signs of its becoming - foreshadowings of the future. Just as on the one side the pre-bourgeois phases appear as *merely historical*, i.e. suspended presuppositions, so do the contemporary conditions of likewise appear as engaged in *suspending themselves* and hence in positing the *historic presuppositions* for a new state of society (Marx, 1973: 460-461).

Marx's work should be used to research particular concrete societies at particular times and places, not as an abstract and ahistorical theory (Cook, 1976-7).

Second, Davis (1980) identifies two instances in Marx's work where labourers own the means of production yet are exploited nonetheless. The first of these appears in *The Eighteenth Brumaire of Louis Bonaparte* (1978). Marx wrote,

> the small holding of the peasant is now only the pretext that allows the capitalist to draw profits, interest and rent from the soil, while leaving the tiller of the soil himself to see how he can extract his wages (Marx in Tucker, 1978: 611).

While it must be acknowledged that Marx's analysis of the French peasantry reveals that the propertied labourer may be exploited by capital, it shows that not all exploitation takes the classic capitalist labour forms. In this instance, Davis (1980: 139) argues, two essential elements are missing. First, a contractual accord between capitalist and labourer which specifies the time and terms for the employment of labour power. Second, capitalist control of the production process. Consequently, 'Marx's analysis of the peasant's situation, then does not entirely satisfy our desire to prove the possibility that propertied labourers may become involved in capitalist relations' (Davis 1980: 139). Marx's analysis of piece-wages does this very precisely.

Piece-wages, wages paid for work performed outside the factory by labourers who own the means of production, give an illusory sense that the recipient stands in a different position to workers in other social modes of production. However,

> the object of his exchange is a direct object of need, not exchange value as such. He does obtain money, it is true, but only in its role as coin; i.e. only as a self-suspending and vanishing mediation. What he obtains from the exchange is therefore not exchange value, not wealth, but a means of subsistence, objects for the preservation of his life, the satisfaction of his needs in general, physical, social etc. It is a specific equivalent in means of subsistence, in objectified labour, measured by the cost of production of his labour. What he gives up is his power to dispose of the latter (Marx, 1973: 282).

Piece-wages therefore represent the transfer of surplus-value from the piece-worker to the capitalist, while ensuring that maintaining the quality and intensity of production is in the personal interest of the labourer (Marx, 1976: 606). This allows the capitalist to increase extraction rates for both relative and absolute surplus value, and to control the entire production process at minimum cost and risk. Consequently, Davis (1980) argues, while the 'independent' producer has been regarded as a barrier to the establishment of capitalist relations, the producer should in fact be regarded as a firm basis for capitalist development. This contention can be borne out in analysis of the advent and subsequent development of contract grape growing in the New Zealand wine industry.

Contract grape growing and the New Zealand wine industry

Contract grape growing existed before 1960 but it only accounted for four percent of production at that time (McDonald, 1973: 65). However, contract growing was largely responsible for the rapid expansion of vineyard acreage between 1965 and 1970. We should not be surprised at this dramatic increase in contract grape growing after 1960. Contract farming accrues considerable benefit for capital, and it was to have important implications for the New Zealand wine industry.

The rise of contract grape growing was driven by the inter-regional expansion of the large Auckland-based wineries. There were various reasons for this expansion. First, spatial diversification offered insurance against localised climatic hazards. Second, there was the unavailability of large contiguous blocks of suitable land, the result of agricultural land in West Auckland being subdivided into 'ten acre blocks' for residential purposes.

Compounding this subdivision problem were significant increases in land values and rates (Townsend, 1976: 84). As a result, in the 1960s both Montana and Corbans, two of New Zealand's largest wineries, established contract relations in Gisborne, and to a lesser degree in the Hawkes Bay (Scott, 1977). From the mid- to late-1970s, medium-sized Auckland wineries like Babich, Delegats, Nobilos and Villa Maria followed suit. Montana and Corbans processed their grapes locally, but the other firms tended to truck grapes from Gisborne to Auckland for processing. An important aspect of the development

Table 17.1 Source of grapes to wine industry 1960-1997, percentage

	1960	1973	1982	1992	1997
Contract grower	4	35	72	75	72
Winemaker	96	65	28	25	28

Source: Workman (1993); WINZ (1997).

of Gisborne as a grape growing region was the rise of the grape variety *Muller Thurgau*, which was ideally suited to the soils and climate of the area. Grapes of consistent quality and high per-acre yields, combined with mechanical harvesting, technological advances in processing and packaging, and consumer demand for bulk table wine, meant that the *Muller Thurgau* variety was well suited to contract growing.

In late 1989, the New Zealand Commerce Commission began an investigation into the wine industry, following allegations that wine companies were colluding in an attempt to fix and hold down the price of grapes. While it was concluded that there was no apparent price fixing amongst wine companies, it was revealed that Montana and Penfolds (part of the Montana group since 1987) were involved in collective price negotiation, which contravened the Commerce Act 1986. The New Zealand Grape Growers' Council sought authorisation from the Commerce Commission to sanction collective negotiation and the fixing of grape prices. The Commission attempted to find out whether the detriments from this practice outweighed any public benefits of the practice, in terms of the lessening of competition.

The Commission prefaced its analysis of the Montana and Penfolds contracts with some general comments regarding grape growing and winemaking. In particular, it noted how volatile these industries were:

> The markets are subject to fluctuations in both price and the quantity and variety of grapes and wine produced. This is due, in part, to the size of the New Zealand industries compared to overseas markets, the impact of climatic conditions, and the fact that consumer preference for different varieties is continually changing. It has also been due to the changes in government policy in respect of the industries which have resulted in higher taxes and lower levels of protection (New Zealand Commerce Commission, 1991: 575).

Consequently, larger wine companies sought to attract growers to the industry by guaranteeing the purchase of grapes. The changing nature of the industries, coupled with the high input costs and the long-term commitment necessary meant that most growers sought the 'relative' security of a contract. Contracts usually specified the price per tonne, the sugar content, the responsibilities relating to harvesting, delivery, crop insurance, time of payment and the passing of risk. Generally, the risks associated with agricultural production remained with the farmer (Smith, 1986: 230).

Prior to 1991 Montana and Penfolds contracts were for periods of 10 and 99 years respectively. Under the contracts, Montana and Penfolds agreed to purchase a specified annual quantity of grapes, of a particular variety or varieties, from a specific area. Montana also had the first option to acquire any grapes above the set volume. Structured as they were, the contracts placed growers in a subordinate position to Montana. In particular, the specification as to the variety to be grown meant that growers had to adhere to Montana's viewpoint as to the regional specificity of grape varieties. Similarly, Montana's right of refusal over surplus grapes from the contract area ensured that in times of oversupply, the perishable product was left in the hands of the grower. To compound matters, an 'act of God' rendered the contract null and void and Montana also reserved the right to reject any grapes which did not meet the criteria of the contract. In part, it was for these reasons that historically the contracts were developed on a regional basis. This reflected the variation between regions in grape variety, and the requirements of growers in that particular region. Consequently, each region had its own group to negotiate with Montana on behalf of all growers. The Grape Pricing Committee, which comprised four representatives from Montana and four from the growers, attempted to determine a base 'price per hectare'. If agreement could not be reached then an independent chairperson was brought in to arbitrate.

> The factors considered in establishing the purchase price per hectare included the price of the previous vintage, changes in national wine stocks relative to sales, and the changing cost of production for growers. In 1985 the comparative price of South Australian grapes was added to this set of criteria (Workman, 1993: 94).

In comparison, the prices negotiated under Penfolds contracts were to be within a 10 percent margin of the gross average price paid by the three largest winemakers in the last vintage.

However, in 1988, due in part to Penfolds being subsumed under the Montana banner, a group comprising representatives of Montana/Penfolds and of each regional growers group was established. The aim was to reach agreement on a new standard contract for all Montana and Penfolds growers. The new formula saw the 'price per hectare' vary according to the grape variety, with no adjustment for sugar content, and prices for grapes above the contracted amount based on the wine market. However, before a resolution could be reached, the Commerce Commission had to decide whether such collective negotiation was allowable under the Commerce Act 1986.

The subsequent submissions to the Commission revealed marked conflict between grape growers and wine companies. The New Zealand Grape

Growers' Council argued that removal of collective negotiation would be detrimental to small independent producers, who did not have the resources to negotiate individually. Owen Jennings, President of Federated Farmers, suggested that this application would have serious implications for the whole rural sector:

> Stable and secure commodity pricing is a public benefit as it ensures the viability of rural communities. Without collectivity, farmers and growers will always be placed in a vulnerable negotiating position, especially when negotiating with monopoly producers (New Zealand Commerce Commission, 1991: 582).

The removal of collective negotiation, it was argued, would allow large wine companies to offer uneconomic prices to growers, thereby removing stability from the industry.

Conversely, wine companies argued that the practice of collective bargaining was detrimental to the industry. First, collective pricing resulted in higher prices being paid to growers.

> To say otherwise is to say that growers' representatives on the Joint Committee do not have an interest in achieving the highest return for growers or that their efforts are ineffective. Neither possibility appears likely (New Zealand Commerce Commission, 1991: 587).

Second, collective pricing would be a recipe for mediocrity:

> The uniformity of prices arising from collectivity reduces the incentive for individual growers to innovate, to improve quality beyond the level required by the contracts or to anticipate changes in the market place. Under collectivity a grower who took these steps would be incurring a risk which, because of price uniformity, could not be rewarded...It is the Commission's view that this imposes a cost on the wine industry which lessens its ability to sell on the international market, and to compete against imports on the domestic market. Any lessening in the competitiveness of the wine industry may ultimately impact on the future viability of the grape growing industry (New Zealand Commerce Commission, 1991: 590).

In its ruling, the Commerce Commission considered that while it was desirable for there to be equitable bargaining power, the fact that Montana was the industry leader did not preclude competition from other wine companies and did not mean that it occupied a monopoly position. Therefore, while the prices determined for Montana's contract growers influenced the price paid by other wine companies, competition ensured that it was not necessary for both parties to have equal bargaining power. Rather, the relative bargaining position of wine companies and growers should reflect the supply/ demand imbalance. Consequently, by 1991, individual price negotiation by Montana saw grape prices fall by as much as 35-50 percent. This was attributed by Montana to four factors. First, grape prices were artificially high due to collective pricing. Second, New Zealand wines had to be price competitive due to greater competition from imported wines, especially from Australia. Third, the 1990 vintage was exceptionally large, leading to an oversupply. Finally, the price of wine needed to drop (via lower grape prices) in order to boost dwindling consumption (New Zealand Commerce Commission, 1991: 579).

According to Workman (1993: 96), the marked drop in prices caused bitter conflict between family grape growers and Montana. Gisborne growers, in particular, were at the forefront of this conflict. In Gisborne, Montana harvested 46 percent of the grape crop, with a corporate/contract production ratio of 10:90 (New Zealand Commerce Commission, 1991: 582). After unsuccessfully challenging the Commerce Commission ruling in court, growers were faced with blatant attacks for their role in opposing Montana.

> One story which indicates the depth of feeling between these groups, concerns Montana's refusal to accept grapes from one of the leaders of the growers on the basis that they were diseased. The grower took the grapes to another wine company, and the wine made from these grapes won a gold medal. When this other company attempted to sell the wine, they were told by an Auckland liquor outlet that Montana threatened to stop supplying the outlet if it stocked the wine (Workman, 1993: 96).

Although Montana lost a number of its contract growers as a result of the drop in grape prices, the company was able to reduce its exposure to the Gisborne region by importing three million litres of bulk wine from Australia in 1992.

Conclusion

The New Zealand Grape Growers Council attempt to have state-sanctioned collective bargaining, as held by wage workers, was ultimately unsuccessful. This meant that growers had to operate in the market as independent entities.

Importantly for the wine companies, this saw not only the price of grapes fall, but also allowed for the extraction of surplus value via both direct and indirect control. Direct control was exerted through the contract which removed the farmer's exclusive control over on-farm decisions. The contract stipulated the variety of grapes to be grown and the amount to be purchased. It gave first right of purchase to the wine company and allowed for the passing of risk from the contractor to the grower. Indirect control was exerted through the maintenance and enforcement of quality standards (which ensured that there were increased viticultural demands), the threat of wine being imported from overseas, and in extreme cases the ostracising of growers and their produce. Consequently, it can be said that the rapid increase in contract grape growing in the New Zealand wine industry has had a number of implications. From our perspective, and in agreement with Davis (1980: 147), the most important of these is that, reduced to a piece-worker, 'the family farmer may direct towards the corporate capitalist an antagonism similar to that which is (periodically) displayed by the industrial worker'.

References

Alavi, H. and Shanin, T. (eds), (1988) 'Introduction to the English Edition: Peasantry and Capitalism', in Kautsky, K., *The Agrarian Question*, Zwan Publications: London, pp. xi-xxxix.

Bernstein, H. (1978-9), 'Concepts for the Analysis of Contemporary Peasants', *The Journal of Peasant Studies*, Vol. 6, pp. 421-443.

Cook, S. (1976-7), 'Beyond the Foremen: Towards a Revised Marxist Theory of Precapitalist Formations and the Transition to Capitalism', *The Journal of Peasant Studies*, Vol. 4, pp. 360-389.

Davis, J. (1980), 'Capitalist Agricultural Development and the Exploitation of the Propertied Laborer', in Buttel, F. and Newby, H. (eds), *The Rural Sociology of the Advanced Societies: Critical Perspectives*, Croom Helm: London, pp. 133-154.

Goodman, D. and Redclift, M. (1985), 'Capitalism, Petty Commodity Production and the Farm Enterprise', *Sociologia Ruralis*, Vol. 25, No. 3/4, pp. 231-247.

Goodman, D., Sorj, B. and Wilkinson, J. (1987), *From Farming to Biotechnology: A Theory of Agro-Industrial Development*, Basil Blackwell: Oxford.

Kautsky, K. (1988), *The Agrarian Question*, Zwan Publications: London.

Little, P. and Watts, M. (eds), (1994), *Living Under Contract: Contract Farming and Agrarian Transformation in Sub-Saharan Africa*, University of Wisconsin Press: Madison.

London, C. (1997), 'Class Relations and Capitalist Development: Subsumption in the Colombian Coffee Industry, 1928-92', *The Journal of Peasant Studies*, Vol. 24, No. 4, pp. 269-295.

Mann, S. (1990), *Agrarian Capitalism in Theory and Practice*, University of North Carolina Press: Chapel Hill.

Marx, K. (1973), *Grundrisse*, Allen Lane: London.

Marx, K. (1976), *Capital (Volume 1)*, Penguin: London.

Marx, K. (1978), 'The Eighteenth Brumaire of Louis Bonaparte', in Tucker R. (ed.), *The Marx-Engels Reader*, W. W. Norton and Company: New York.

McDonald, T. (1973), 'The New Zealand Wine Industry', *News Media Forum of the New Zealand Liquor Industry Council*, N.Z. Licensee: Wellington.

Montana Wines Ltd. (1979), *The Montana Story*, Montana Wines: Auckland.

New Zealand Commerce Commission (1991), *New Zealand Commerce Decisions, Volume 2 1989-1991*, CCH New Zealand Limited: Auckland.

Scott, D. (1977), *A Stake in the Country*, Southern Cross Books: Auckland.

Smith, W. (1986), 'Agricultural Marketing and Distribution', in Pacione M. (ed.), *Progress in Agricultural Geography*, Croom Helm: London, pp. 219-238.

Stoler, A. (1987), 'Sumatran Transistions: Colonial Capitalism and Theories of Subsumption', *International Social Science Journal*, Vol. 114, pp. 543-562.

Townsend, M. (1976), *Location of Viticulture in New Zealand*, Unpublished MA thesis, University of Auckland: Auckland.

Watts, M. (1996), 'Development III: The Global Agro-food System and Late Twentieth-century Development (or Kautsky *redux*)', *Progress in Human Geography*, Vol. 20, No. 2, pp. 230-245.

Wine Institute of New Zealand (1997), *Annual Vintage Survey*, Auckland.

Workman, M. (1993), *Geographic Organisation of the Wine Industry in New Zealand*, Unpublished MA thesis, University of Auckland: Auckland.

18 An Action Learning approach to grower-focussed change: Research among cotton producers in Queensland

GEOFFREY LAWRENCE, MELISSA MEYERS, STEWART LOCKIE
AND RICHARD CLARK

Introduction

Irrigation industries in Queensland currently consume approximately two million megalitres of water per year, or some 65 percent of all water used in the state (Queensland Irrigation Council Steering Committee, 1997).[1] The recognition of water as a valuable and scarce resource has prompted research and development agencies such as the Land and Water Resources Research and Development Corporation (LWRRDC), to devote considerable attention to investigating extension models and processes which may lead to more sustainable management of water resources. One investigation - the 'local best practices' (LBP) project - was initiated in 1995 as one of five pilot projects funded by LWRRDC under the National Program for Irrigation Research and Development. The purpose of four of the five pilot projects is to assess the participatory action management (PAM) model of extension (Chamala and Keith, 1995) in irrigation industries across Australia. For our particular project, we implemented a slightly different approach, called the 'local best practices' process, developed and trialled in the beef industry by Richard Clark of the Queensland Department of Primary Industries (DPI).

In deciding on cotton production, we chose a 'big' irrigation industry, that is one which, although a heavy user of water, contributes up to A$1 billion to the Australian economy annually (*Cotton Insight*, May 1996: 3). Our study focusses on the Emerald Irrigation Area (EIA) in Central Queensland, an area producing around 140,000 bales of cotton and earning approximately A$70 million dollars each year (*Central Queensland News*, 28 May 1997). The most striking feature of the irrigated cotton industry is the intensive nature of the production system which is characterised by a high level of chemical inputs and high risk capital outlays, with yield and profit dependent on a guaranteed water supply and technological 'solutions' to production problems.

Unlike many broadacre farmers who believe they are on a 'technological treadmill' and who believe they have little choice other than to use available chemical and other inputs in an effort to remain viable in an increasingly competitive market place (see Lawrence, 1987; Gray *et al.*, 1995; Lockie *et al.*, 1995; Lockie, 1996), cotton producers eagerly seek new technologies to solve current production problems and to maintain profitability. The cotton industry promotes the benefits of adopting a techno-strategic approach to farm management. An Executive Director of Cotton Australia remarked in 1997 that the 'industry's wholehearted commitment to adopting science and technology on farm was largely responsible for this year's record harvest', affirming the notion that to be progressive, successful and of benefit to the industry, utilisation of output-boosting technology will continue to be essential (*Central Queensland News*, 5 July 1997).

Despite the perceived advantages of current approaches, there are serious concerns in Emerald about farm chemicals in the environment, and their impact upon community health (*Central Queensland News*, 2 February 1996; 9 February 1996; 14 February 1996). These parallel more general concerns about chemicals in Australian agriculture (see Short, 1994; McHugh, 1996). There is also community concern in Emerald about the sustainability of an enterprise that uses water so intensively in a semi-arid region. Water restrictions in the 1995/96 season and a zero allocation at the beginning of the 1996/97 season confirmed local community suspicions that the current demands on existing water resources may not be sustainable in the long-term (*Cotton Insight*, May 1996: 13). These protests, however, are fleeting and lack the coordination and intensity which might impact on the industry in the region. Agencies servicing the industry have become well versed in promoting the benefits of the industry, local employment and wealth creation, and minimising the perceived impacts - such as implementing chemical application guidelines which encourage 'communication' with nearby residents.

In terms of traditional extension activities, the local cotton industry is well serviced by industry-based agencies such as the Cotton Research and Development Corporation, the Australian Cotton Research Institute, and State government agencies such as the Department of Primary Industries and the Department of Natural Resources. Within the 'transfer of technology' (TOT) philosophy, administered by the above agencies, farmer 'participation' could more usefully be described as 'consultation'. Farmer involvement in extension activities within a 'top-down' paradigm is limited to attendance at field days, providing an on-farm location for trials and 'token' representation on various agency boards, committees and management groups. At the beginning of the local best practices (LBP) project there was an absence of any 'bottom-up' participatory approaches in the Emerald cotton industry.

Local best practices

The LBP approach is a participative group-extension method which is based on a 'farmer first' or 'bottom-up' paradigm (see Chambers, 1989). This broad philosophy of farmer 'participation' can be seen as a response to the perceived inadequacies of the traditional TOT model of extension, which has marginalised local knowledge, and taken an uncritical technocratic approach to solving production (and to a lesser extent, environmental) problems. There is currently something of a 'crisis of extension' in Australia which is marked by a reduction in publicly-funded activities, concern with the efficacy of older traditional forms of extension, and the inability of extension to 'correct' severe environmental degradation (see Chamala and Keith, 1995; Vanclay and Lawrence, 1995). This crisis has been a catalyst to the development of new 'bottom-up' models which seek the development of an alternative paradigm for extension, which question the hegemony of scientific rationality, and which aim to promote other 'ways of knowing' and 'alternative voices' (see Kloppenburg, 1991; Flora, 1992; Vanclay and Lawrence, 1995).

The LBP method of extension provides a structured mechanism to encourage producers to develop and utilise local knowledge (see Figure 18.1). Where a TOT adoption-diffusion approach would aim to persuade producers to adopt technological or scientifically-based *improvements*, LBP aims to generate ideas, problems and opportunities for improvement from a producer perspective. Where the TOT top-down paradigm views producers as 'needing' the inputs of science (and finding means of having them adopt technologies developed by researchers), LBP recognises the wealth of experience and knowledge of producers and aims to provide a process for producers to reflect on their current management practices and to attempt to improve their local situation (Meyers, 1997). In doing so, LBP brings expert advice to the group - as demanded by the group. It utilises what Clark (1996) has described as a 'specialist questioning technique', involving specialists with information on local management practices and property characteristics. These specialists

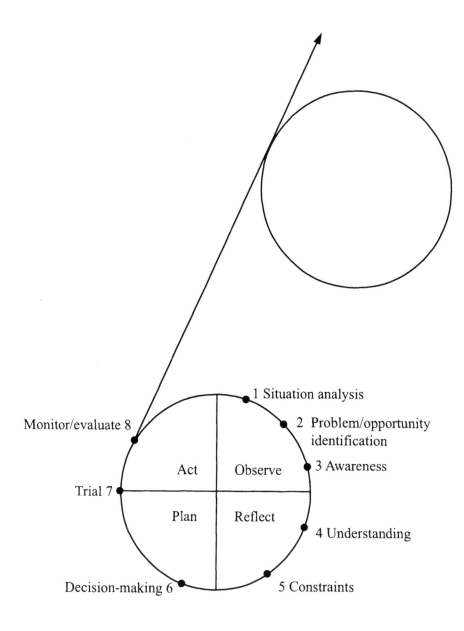

Figure 18.1 Action learning cycle - local best practice

are asked to contribute ideas and potential solutions to the problems raised by producers. In this sense, it is science which serves the growers, not the growers who mechanically 'adopt' new science delivered from on-high by the 'experts'. Finally, the method is based upon acknowledgment that problems are often specific to local areas - that they arise from quite unique combinations of land forms, vegetation types, proximity to water and so forth, and that attempts by extension or other advisers to impose 'solutions' which may have worked elsewhere, are likely to fail. The utilisation of local knowledge to solve local problems is at the heart of the LBP model.

The approach works by establishing small groups of producers in a given local area for the purposes of working towards improved practices and sustainability. Working in small groups, with a trained facilitator, producers are encouraged to:

> describe current knowledge and perceptions of 'Best Practices', describe problems, solutions, opportunities and constraints in the context of improving current practices and learn from one another about current practices and opportunities to improve these (Clark *et al.* 1996: 92).

The approach to the cotton industry in Emerald was initially made through the local Cotton Growers and Irrigators Association and resulted in the formation of one group of approximately ten growers. The Emerald Irrigation Area is in Central Queensland, some 300 kms west of Rockhampton. It is separated into two 'regions' by the Nogoa River, which seems to have resulted in a social - as well as a geographical - division within the grower population. The eastern side consists of relatively new cotton farms owned by younger growers, some new to cotton, who tend to participate heavily in the industry; the western side of the river consists of older cotton farms, with an ageing grower population, some of whom have been on the same cotton property for thirty years. The first group was formed by explaining the process and its potential benefits to members of the growers' association. A member of the growers' association agreed to nominate and contact members of a 'local' group. The group formed consisted of growers located on the eastern side of the river, primarily due to the dominance of these growers in the association. Several attempts at forming additional groups in the early parts of the project were unsuccessful. Many growers said they were either too busy, or had no interest in group processes. A more recent attempt has resulted in the formation of a second group (and potentially a third) on the western side of the river. Their formation is primarily due to assistance from local DPI extension staff and the perceived 'success' of the first group.

The first group (The Weemah 'best practices' group) utilised a problem solving cycle to move from an analysis of the current situation, through to identifying local problems, opportunities and constraints. Considerable time was spent with this group analysing the range of management practices. Due to the delay in the formation of the second group (The Wills Road 'best practices' group), a revised version of the problem solving cycle was developed, with a defined focus on irrigation practices. This focus has allowed the group to move quickly to identifying problems and possible opportunities. Both groups have developed critical and insightful questions regarding current water management practices, and generated innovative ideas to improve irrigation management. The Weemah group identified drip irrigation technology as having the greatest potential for improved irrigation efficiency. The idea was dismissed as unfeasible and inappropriate by all of the relevant research, development and extension agencies in the local (and wider) cotton industry. The group has persisted in researching the technology through the monitoring of established sites in the local region, and visiting other sites which use the technology throughout Australia's cotton growing areas. The results from the monitoring being undertaken by the group, have created some renewed industry interest in drip irrigation. Both groups were involved in the organisation of a workshop with invited researchers, designed to explore issues relating to irrigation efficiency (in terms of farm layout and reticulation systems), nitrogen use (minimising losses associated with flood irrigation) and the use of polyacrilamides to minimise soil erosion. The focus of the workshop was on working together with researchers to develop action plans, based on relevant local information, which can be implemented on individual farms.

To date, the project has been successful in motivating producers to examine current management practices and to explore potential opportunities for improvement. There is some evidence to suggest that this will lead to change in practices, with many of the growers giving serious consideration to alternative irrigation practices which may include drip irrigation systems, or undertaking further work on maximising the water-use efficiency of flood irrigation. But the extent to which this type of approach can promote change and self-reliance, is limited by the current research and extension environment, which promotes the scientific enterprise and marginalises producer knowledge.

Trying circumstances

One of the major constraints on the progress of the groups has been a lack of support (and understanding) from among research, development and extension (RD&E) agencies. Despite the substantial rejection of the TOT model and the emergence of new participatory approaches in extension, the 'top-down' paradigm still prevails in the current RD&E structure (Chambers, 1989; Woods *et al.*, 1990; Clark *et al.*, 1996). Given that the underlying assumptions of TOT and the newer 'bottom-up' approaches are fundamentally different, it is perhaps not surprising to find the RD&E agencies struggling to incorporate and support participative approaches. The funding of trials and activities is largely limited to the defined 'priorities' of particular agencies, making it difficult to obtain funding for local group-driven initiatives. In this sense the groups remain marginal with a low level of recognition and support. Lockie (1995) suggests that in the Landcare model, it is necessary for groups to supplement government funding with additional resources from the wider community. In a similar way to the LBP project, Landcare groups are provided with insufficient funding to change dramatically existing practices, without seeking support from other avenues. In seeking additional resources, Landcare groups have the advantage of a high level of public exposure and wider community backing. Groups, such as the established LBP groups, are less likely to be successful in attracting support for what is, essentially, a project to improve *individual* farm operations.

Adequate support and resources are critical to the success of group approaches. The LBP groups have proven to be labour intensive in the early stages of group formation, requiring organisation of the group and facilitation through problem solving cycles. They are also resource intensive in the latter stages, requiring technical knowledge and skills, information and - importantly - financial support for trials. Yet governments are seeking to rationalise their activities and to withdraw, where possible, from the provision of what they consider should be 'user pays' services. The current view of most governments at the state and federal levels is that if a service is needed by growers it should be provided by private enterprise, and paid for by the recipient. In spite of this, farmers *do* want state-based extension, and tend not to pay for alternatives when it is withdrawn (see Vanclay and Lawrence, 1995). A cynical, but by no means inappropriate way of understanding the current government interest in farmer-driven approaches, is to view them as a vehicle for the continued reduction of traditional state support for agriculture.

If this is the case, and governments do not adequately resource group-based projects, it is likely that group extension will prove to be as spectacular failure as any alternative model of extension. In terms of state commitment to agriculture, a move toward group-based extension may lead to an *increase* in demand for individual extension services, with groups of growers being motivated to explore local problems and opportunities and attempt to implement changes on-farm (see Vanclay and Lawrence, 1995). An argument could be mounted that, given the importance of group-based problem solving, governments might simply fund grower groups to 'purchase' services and advice as they need it, possibly with funding drawn from existing extension or other activities. While some of the producers within the LBP group have shown high levels of commitment and enthusiasm in participation, and recognise the virtues of self-direction, they believe they do not have the skills and/or the time to organise meetings. Their time is precious and they believe it is someone else's job to organise and facilitate.

The group has expressed its desire to receive continued support and direction from facilitators. This will, in effect, mean that the initial 'start-up' support required for the group will need to continue and may be a constant critical factor in the success of these groups. Other projects investigating participatory group approaches in agriculture have found that a committed facilitator or 'champion' within the group is critical to the continued progress of that group (see Kernot *et al.*, 1996; Clark *et al.*, 1997). Referring back to the earlier point regarding government funding, the question here becomes one of 'who pays' for this continued support of local groups?

Gaining widespread producer participation in the LBP project has also proven to be a challenge. The sheer number of extension activities in the cotton industry has led to competition between projects for farmer involvement. Producers in the LBP groups have indicated that they feel overwhelmed by the number of events and meetings that are scheduled, particularly through the growing season. This 'overload' of contact and time spent on extension activities (together with the more obvious feature of 'information overload') means that the LBP project becomes one of many options, and may be seen as less important than, say, a workshop or shed meeting, which have an established place in farmer commitments to extension. The project, to date, has attracted the participation of growers who already participate heavily in industry extension. A major criticism of the traditional TOT model of diffusion/adoption relates to the inability of the model to cater adequately for (or service) all levels of farmers. The assumption of the TOT diffusion/adoption model is that if extension targets 'innovators', the resulting knowledge and information will filter through (or trickle down) to the rest of the 'target' population

(Vanclay and Lawrence, 1995: 27). This concept has proved to be less than successful, and it seems that the TOT approach benefited the 'progressive' or 'top-end' farmers, while doing little for the so-called 'laggards' or 'bottom-end' farmers. In theory the LBP approach offers participation to all producers. In practice, however, current group members could all be considered to be 'innovators' or 'progressive' farmers, with a high level of participation in other extension activities and in industry bodies, a reasonably high level of education, a successful farming enterprise, and a willingness to experiment. A major challenge for the LBP approach lies in its ability to attract other farmers, and in particular those who do not participate in other traditional extension activities and industry events.

There is also a noticeable absence of participation by women farmers, despite attempts to include them from the outset. Lockie (1995) suggests that this lack of involvement by women is a common phenomenon in the membership of farm organisations - one related to prevailing and entrenched ideologies such as that of patriarchy. Further, there are indications that within established Landcare groups, the active involvement of women actually leads to more effective group results (see Lockie, 1995). The non-participation of women in the LBP groups may perpetuate the concept of men as 'farmers' and women as 'wives' and negate a valuable contribution, i.e. the 'different voice', of women in the farming enterprise.

An alternative form of extension?

Participative, 'bottom-up' approaches such as the LBP method raise some interesting questions about the role of extension and the nature of knowledge and, most importantly, call into question the hegemony of the scientific paradigm.

The role of public extension needs to be carefully considered. Traditional extension services have been designed as a way to 'help' farmers to become more productive. These services have been heavily dependent on science and technology to provide universal solutions to diverse situations. We now know that this has been less than successful, with extension services often criticised for being irrelevant or impractical in terms of on-farm problems. Flora (1992) argues that the traditional TOT approach has assumed farmers are 'interchangeable', with knowledge of local conditions becoming less important than following prescribed best management practices (see also Kloppenburg, 1991). Kloppenburg (1991) and Molnar *et al.* (1992) agree that first, there is a fundamental difference between the perspective of the scientist and that of the farmer, and second, that while operating in fundamentally different ways, the two perspectives are indeed complementary. The role of the scientist is to produce generalisable solutions to problems in a controlled environment, while the role of the farmer is to attempt to maximise efficiency, including solving problems, within an ever-changing and specifically situated locality (Molnar *et al.*, 1992: 86). What appears to have been missing, and seems to form the basis of Kloppenburg's (1991) critique of traditional approaches, is a recognition of the importance of the role of the farmer in the process of change - of the validity of 'local voices' in that change. What has been missing from the system is the critical component which encourages interaction between the scientist and the farmer – with the farmer being able to call upon specific aspects of the work of 'experts' which the farmer identifies as essential to his/her future activities. Another missing feature has been the benchmarking of activities by farmers. It is important for them to understand how their changed on-farm practices have led to different outcomes. With TOT and some group extension practices nothing is ever measured by the producers themselves; the 'success' or otherwise of new practices is, instead, crudely measured by the increased profitability of the enterprise, an outcome which might have more to do with world market conditions than with the efficacy of changes at the farm level. They need to be empowered to undertake 'benchmarking' - something which is, for most, a new activity.

Kloppenburg argues that it is impossible to incorporate farmer voices into the current system where scientific knowledge is valued above all else, and limits the contribution of alternative perspectives. What is required is a reconstruction of the paradigm of agriculture that articulates multiple ways of knowing (Kloppenburg, 1991). Kloppenburg (1992) argues that the problem with the dominance of scientific knowledge in the agricultural production system, is not that it is 'invalid' or 'incorrect', but that it is a *partial* account, just as other forms of knowledge are partial. He suggests that to move closer towards an holistic understanding of farming systems, we need to engage with and encourage 'conversation' between various partial accounts (1991: 536). It could be that local producer groups have a role to play in developing local knowledges which have been previously marginalised and devalued. So what is required, according to Kloppenburg, is not the replacement of 'one hegemonic epistemology' with another, nor the 'blind promotion of farmers' knowledge to a category of superior status', but a deconstruction of the existing paradigm which promotes 'scientific' knowledge, and a reconstruction of a new paradigm which allows the multiplicity of voices to be recognised and valued (Kloppenburg, 1991: 542; 1992: 102; Molnar *et al.*, 1992: 86).

As a way of encouraging the development of local technical knowledge, the LBP approach holds great promise. The method provides a structured way to undertake the reconstruction of local knowledge, through the formation of locally-based groups aimed at exploring the local situation and generating problems and opportunities from a producer perspective. In an alternative paradigm, which encourages an interplay between the removed world of science and technology and the situated localised world of the farm, LBP could be effective in mediating and merging the 'accounts' or 'knowledges' of science with those of farmers.

In many ways a new paradigm which encourages a combination of approaches, eliminates many of the criticisms of the 'bottom-up' models, and indeed, many of those related to the TOT model. This is because most of these criticisms reflect the inadequacy of relying on one 'partial' account of the system. Hence, any criticism which suggests that 'bottom-up' approaches rely too heavily on producer awareness of problems which may be difficult to recognise without technical assistance, or which producers may simply choose to ignore (particularly environmental problems) (see Martin and Lockie, 1993). It is argued, too, that producers may not have the necessary skills and expertise to solve complex problems (a criticism which can be answered through interaction between producer and specialist). Similarly, criticisms that the TOT model provides information which is irrelevant and impractical in local situations, and that it uncritically promotes the products of science and technology may be countered by integrating producer knowledge into the system.

While this appears a simple solution to the 'crisis in extension', Flora (1992) warns us that the de/reconstruction project will not be easy. The hegemony of science is not incidental, and not limited to agricultural discourse (see Bridgstock *et al.*, 1998; Hindmarsh *et al.*, 1998). The power of science and technology in our culture is widespread and persuasive. Flora (1992) argues that few, if any, of the current agricultural institutions are capable, or even willing, to alter drastically their current scientific 'way of knowing'. Such a step would alter the distribution of power, to the detriment of those currently having the most influence.

Local best practices and agri-food restructuring

A major criticism of the contemporary agri-food restructuring literature is that it veers uncomfortably close to structural determinism. Theories are often pitched exclusively at the macro-level and so miss, or deliberately ignore, the complexity of on-farm activity – including local responses to global processes. For Buttel (1996: 33), it is necessary to find ways of understanding local configurations and to see how these might challenge the 'technologically-driven homogenisation of agriculture'. Importantly, for Buttel, agricultural practices and local social conditions and structures are crucial to any understanding of new trends and tendencies, including the ways farmers are addressing or intend to address environmental problems. We need to evaluate the LBP approach both in terms of its potential to motivate a group of producers to alter current trajectories of farming, and in terms of what it reveals about wider agri-food restructuring.

Changes within the group

After two years working with cotton producers, very little has changed in terms of cotton-growing practice. The potential for change is there, of course, but it is clear that, despite attempts by the growers to develop different approaches such as trickle irrigation, they have reverted to existing practices to produce their crops. This raises the issue of what 'lead in' time is necessary before knowledge gained in meetings is translated into on-farm practices. What *has* happened at this stage, according to Meyers (1997: 7-8), is that growers in the LBP group in Emerald:

- have accepted that working in a group situation where knowledge is shared and where they have been focussed on problems which they, themselves, have identified, is a desirable basis for interaction;
- believe that they now better understand the context of local issues and problems in the district (both on- and off-farm);
- have experimented with trickle irrigation, and have found that, compared to current furrow irrigation practices, it reduces water usage by around 25 percent and leads to production increases of up to two bales per hectare in some areas. Such initial 'success' has spurred on the group to demand agencies to assist them in their trials;
- have become more self-reliant and confident in dealing with complex issues, including that of the development of a sustainable cotton production system;
- have been able to compare and contrast the varying practices identified by group members, and in the process have proved incorrect the

prevailing view that all growers were operating in much the same manner.

In this context, on-farm restructuring appears to be driven by an acceptance that there are 'better' (read more profitable and more environmentally-sound) ways of producing agricultural commodities. Group processes foster a cooperative and open approach to problem solving, and encourage group 'ownership'. The question remains, however, as to whether the group is capable of moving away from the more destructive, or environmentally-unfriendly, practices, which are the current basis of profit-making in the industry. Does the LBP model really challenge the hegemony of the scientific paradigm, or does it conform to it?

Kernot *et al.* (1996) argue that utilising traditional extension approaches for specialist input within the LBP model is inconsistent with the producer-driven process which claims to value farmer wisdom and local knowledge. They suggest that involving specialists in workshops could potentially represent a return to conventional extension techniques that devalue the knowledge of farmers. The role and function of specialists needs to be carefully considered and negotiated within any producer-driven model.

Might it also be the case that, while subverting the TOT approach, the LBP model provides support for the maintenance of the 'productivist' paradigm - albeit, in slightly modified form? At the outset, the LBP approach concentrates the minds of producers on immediate problems rather than with visions of the future. The notion here is that if problems are addressed, and solutions found, the process will be viewed as having legitimacy, leading to further 'rounds' of discussion and decision-making about the future.

The larger questions (should we be growing cotton in this area? Are there alternative land use options? What are the impacts of our on-farm practices on wider catchment management?) usually only occur after the successful completion of several action learning cycles. This places responsibility on the facilitator to ensure that the cycles are completed - otherwise the groups' activities may simply be tinkering at the edges of the productivist paradigm.

The lack of involvement of women raises some concerns. There is anecdotal evidence suggesting that women are often more interested in environmental and farm health issues than men. If this is so, their 'exclusion' from group decision-making may skew discussions away from those issues. We simply do not know what influence women may have had, had they been involved in discussions. It might be desirable to form a women's group as part of any development and future promulgation of the LBP approach.

Wider agri-food restructuring

State agencies have had a pervasive influence upon agriculture in Australia, underwriting farm profitability, providing research, supplying quarantine services and, of course, providing extension (Lawrence, 1987). In recent times there has been a withdrawal of the state from many supportive activities, in line with the wider tendency towards a reduced role for government under the influence of economic rationalism (see Rees, *et al.*, 1993; Stewart, 1994; Lawrence, 1996). What remains of RD&E services to agriculture (and particularly extension) is viewed as being very limited. What this study has found is that there is little 'flexibility' within the bureaucracy which would enable it to meet the needs of the LBP groups. Agency officials are simply too busy with existing programs to act quickly (or at all) in providing the expert advice which is an essential input in the LBP model. If producers are developing innovative alternatives, if they require expert guidance, and if that support is not forthcoming, grower frustration will inevitably arise (see Meyers, 1997). Moreover, the state may come to be viewed as a barrier to progress, particularly in those instances in which grower's suggestions support some more environmentally-sound options. Lack of support would seem to be a major hindrance to change. It has been reported that state officials feel 'threatened' by the new approach (Meyers, 1997).

Should the producers be expected to provide their own funds for advice, as they did in this study? Or should governments respond by reorganising their extension practices in order to assist the development of the groups? It should be remembered that the cotton producers are a wealthy group; will other groups of producers - with little financial reserves - be willing or able to fund investigations into new options, and then to actually implement them if they are found to be desirable? If not, what will this mean in terms of any progressive moves toward a more profitable and sustainable agriculture?

The sharing of knowledge throughout the farming community is another issue relating to agri-food restructuring. Is it sufficient to have a group of producers undertake actions which they hope will solve their problems, when others - excluded from the group - are unaware of or are distant from, the decision-making processes? Notions of elitism must be considered here (see Meyers, 1997). If the group comprises the very people who are active in all local meetings and events, is the group simply another vehicle for the 'innovators' to develop, while doing little for those remaining?

In terms of agricultural restructuring, this may mean that the polarisation which is occurring in agriculture between the larger and more capital intensive producers and the smaller/less capital intensive producers, will be promoted by LBP. Or will the polarisation relate to a division between the 'knowledge rich' and the 'knowledge poor' (Jones, 1982) in agriculture? If either of these outcomes occurs, it is likely to result in the real problems of contemporary agriculture - the need to provide economic security and overcome environmental degradation - being ignored. In such circumstances, the potential for LBP to 'move' farming toward more sustainable forms of production will be very limited.

Conclusion

The local best practices process of participatory problem solving has the potential to encourage the development of an understanding of management practices, critical thinking skills and problem-solving techniques, and to contribute to the reconstruction of local knowledge systems. The strength of the process lies in its ability to generate enthusiasm and participation by producers in a process designed to allow them to play a greater role in terms of identifying, understanding and addressing current problems. The process has the potential to generate producer knowledge regarding local problems and issues and to make a worthwhile contribution of ideas and concepts to the RD&E system. However, the applicability and benefits of the approach are limited within the prevailing paradigm, which maintains the sanctity of science above the local technical knowledge of farmers.

In terms of agri-food restructuring, the application of LBP approaches has the potential fundamentally to alter the trajectory of agriculture. However, if support mechanisms are not provided, if women continue to be excluded, if the groups only recognise the productivity-raising side of new approaches to farming, and if an 'elite' emerges whose knowledge cannot be 'generalised' within the farming community, the likelihood of success will be consequently diminished.

Note

1   Research presented herein was made possible through support from the Land and Water Resources Research and Development Corporation.

References

Bridgstock, M., Burch, D., Forge, J., Laurent, J. and Lowe, I. (1998), *Science, Technology and Society: An Introduction*, Cambridge: Melbourne.

Central Queensland News, 2 February 1996; 9 February 1996; 14 February 1996; 28 May 1997; 5 July 1997.

Chamala, S. and Keith, K. (1995), *Participative Approaches for Landcare: Perspectives, Policies, Programs*, Australian Academic Press: Brisbane.

Chambers, R. (1989), 'Reversals, Institutions and Change', in Chambers, R., Pacey, A. and Thrupp, L. (eds), *Farmer First: Farmer Innovation and Agricultural Research*, Intermediate Technology Publications: London.

Clark, R., Bourne, G., Cheffins, R., Esdale, C., Filet, P., Gillespie, R. and Graham, T. (1996), *Sustainable Beef Production Systems Project: Beyond Awareness to Continuous Improvement*, Department of Primary Industries: Rockhampton.

Flora, C. (1992), 'Reconstructing Agriculture: The Case for Local Knowledge', *Rural Sociology*, Vol. 57, No. 1, pp. 92-97.

Gray, I., Phillips, E., Ampt, P. and Dunn, A. (1995), 'Awareness or Beguilement? Farmers' Perceptions of Change', in Share, P. (ed.), *Communication and Culture in Rural Areas*, Centre for Rural Social Research, Charles Sturt University: Wagga Wagga, pp. 53-69.

Hindmarsh, R., Lawrence, G. and Norton, J. (1998), *Altered Genes: Reconstructing Nature*, Allen and Unwin: Sydney.

Insight (1996), South Queensland Cotton Growers Association Incorporated, May.

Jones, B. (1982), *Sleepers, Wake!*, Oxford University Press: Melbourne.

Kloppenburg, J. (1991), 'Social Theory and the De/Reconstruction of Agricultural Science: Local Knowledge for an Alternative Agriculture', *Rural Sociology*, Vol. 56, No. 4, pp. 519-548.

Kloppenburg, J. (1992), 'Science in Agriculture: A Reply to Molnar, Duffy, Cummins, and Van Santen and to Flora', *Rural Sociology*, Vol. 57, No. 1, pp. 98-107.

Lawrence, G. (1987), *Capitalism and the Countryside: The Rural Crisis in Australia*, Pluto: Sydney.

Lawrence, G. (1996), 'Contemporary Agri-food Restructuring: Australia and New Zealand', in Burch, D., Rickson, R. and Lawrence, G. (eds), *Globalization and Agri-food Restructuring: Perspectives from the Australasia Region*, Avebury: Aldershot, pp. 45-72.

Lockie, S. (1995), 'Rural Gender Relations and Landcare', in Vanclay, F. (ed.), *With a Rural Focus*, Proceedings of the Rural Section, Australian Sociological Association Conference, December 1994, Centre for Rural Social Research, Charles Sturt University: Wagga Wagga.

Lockie, S. (1996), 'Chemical Risk and the Self-calculating Farmer: Diffuse Chemical Use in Australian Broadacre Farming Systems', *Current Sociology: Trend Report on 'Technological Disasters and Community Transformation*, November.

Martin, P. and Lockie, S. (1993), 'Environmental Information for Total Catchment Management: Incorporating Local Knowledge', *Australian Geographer*, Vol. 24, No. 1, pp. 75-85.

Meyers, M. (1997), *Evaluating the 'Local Best Practices' Method of Participatory Problem Solving*, unpublished paper, Rural Extension Centre: Rockhampton.

Molnar, J., Duffy, P, Cummins, K. and Van Santen, E. (1992), 'Agricultural Science and Agricultural Counterculture: Paradigms in Search of a Future', *Rural Sociology*, Vol. 57, No. 1, pp. 83-91.

Queensland Irrigation Technology Steering Committee (1997), *Charter of Operations*, Department of Primary Indiustries: Rockhampton.

Rees, S., Rodley, G. and Stilwell, F. (eds), (1993), *Beyond the Market: Alternatives to Economic Rationalism*, Pluto: Sydney.

Stewart, J. (1994), *The Lie of the Level Playing Field: Industry Policy and Australia's Future*, Text Publishing: Melbourne.

Vanclay, F. and Lawrence, G. (1995), *The Environmental Imperative: Eco-Social Concerns For Australian Agriculture*, Central Queensland University Press: Rockhampton.

Woods, E., Moll, G., Coutts, J., Clark, R. and Irvin, C. (1993), *Information Exchange: A Review*, Land and Water Resources Research and Development Corporation: Canberra.

19 The discourse of sustainable development and the Australian sugar industry: A preliminary analysis

DAVID GRASBY

It is, by now, a commonplace that sustainable development can mean
essentially whatever you want it to mean. The phrase has become a
slogan, truism, shibboleth, cliche, and, for a considerable body of
agnostic opinion, an oxymoron. (Goodman, 1993)

Introduction

This chapter explores the origin of 'sustainable development' and the way in
which 'sustainability' is used as discourse within the Australian sugar industry.
Using the report of the World Commission on Environment and Development
(WCED,1987; commonly referred to as the Brundtland report) as the defining
statement on the topic, this chapter will examine how the notion of sustainable
development has been incorporated into the production of sugar in Australia.
As Goodman (1993) so ably enunciates above, sustainable development can
be interpreted in numerous ways, all of which depend, to a large extent, on
the interests of the groups who call for development to be made sustainable.
As it is understood currently, sustainability invariably refers to either ecological
or economic sustainability. In the case of sugar, the notion of sustainability is
used by organisations which represent the varied interests of the sugar industry,
to direct canegrowers towards practices which attempt to balance the, at times
contradictory, goals of economic growth and ecological sustainability. These
may be worthy goals, but the idea of sustainable development as embracing
purely economic and ecological interests, fails to reflect adequately the intent
of the World Commission on Environment and Development. The Brundtland

report appeals for a more 'holistic' interpretation which gives equal consideration to economic growth, ecological sustainability, and the sustainability of human communities by the alleviation of poverty.

The Australian sugar industry is one enterprise which has been criticised for the damage that the production of sugarcane causes. The effects of the industry on the environment, and on the lifestyles of an increasing number of 'urbanised' rural residents, have led to calls for the industry to become more environmentally sensitive and socially aware. However, sugarcane growers have come under pressure to increase production in order to remain economically viable. Sustainability, therefore, has come to have various meanings and implications, which depend upon the viewpoint taken, and the perspective adopted, by different social actors. Using the literature produced by a number of organisations involved in the sugar industry, this chapter will explore how the concept of sustainability is articulated, and will compare this discourse with the idea of sustainability as framed in the Brundtland report.

The origin of sustainable development

The desire for sustainability is implicit in all works of the pre-Enlightenment era which deal with the notion of 'development'. From the early Greek philosophers onwards, a major concern was to create a form of society which was self-perpetuating (Cowen and Shenton, 1996; Nisbet, 1980). The developmentalist ideas implicit in the works of Plato, Aristotle, Augustine, Sir Thomas More and others, are imbued with the desire to create a society which is self-perpetuating. In the Platonic conception, as outlined in *The Republic*, the ideal society would be created though the application of philosophy to politics. In words which have as much veracity today as they did in his own time, Plato expressed the view that,

> the human race would never see the end of trouble until true lovers of wisdom should come to hold political power, or the holders of political power should, by some divine appointment, become true lovers of wisdom (quoted in Bambrough, 1976: xi).

To Augustine, religion, or to be more precise, the teachings of Jesus, provide the means for the development of a society which, through continuous cycles of growth and decay, would endure in perpetuity. While the Platonic and Augustinian perceptions differ in their approaches to the formation of the 'ideal' society, the essential similarity is that they portray a vision for a community in which the welfare of the citizens is accorded a higher priority than that of the natural environment.

With the arrival of the Enlightenment and the Industrial Revolution in the eighteenth and nineteenth centuries, came an awareness that the process of industrial, or capitalist, development carried with it certain detrimental outcomes. Writers such as August Comte, began to argue for a form of development which, with the aid of scientific principles, would arrest some of the tendencies toward social disorder and environmental damage which had arisen from industrial development. John Stuart Mill also harboured grave misgivings about the course of modern development. In *The Principles of Political Economy*, published in 1848, Mill articulated a vision of sustainable development which not only called for the alleviation of the social evils of development but also the environmental damage which industrial development had, even in its earliest stages, given rise to. Mill stated that:

> If the Earth must lose that great portion of its pleasantness which it owes to things that the unlimited increase in wealth and population would extirpate from it, for the mere purpose of enabling it to support a large, but not a better or happier population, I sincerely hope, for the sake of posterity, that they will be content to be stationary, long before necessity compels them to it (Mill, cited in Cowen and Shenton, 1996: 41).

In that one passage, John Stuart Mill outlined many of the apprehensions which now preoccupy contemporary commentators who argue for a more ecologically sustainable form of development. It is arguable, therefore, that the *modern* idea of sustainable development was conceived in the light of circumstances associated with the development of capitalism which were actually occurring in the nineteenth century. In Mill's notion of 'stationary development', can be discerned an emerging ecological consciousness, and a growing unease with the course that capitalist development had begun to take. To Mill, there was a need to encourage a form of development which would not be detrimental to the natural environment, would benefit the population in its totality, and was capable of being perpetuated over time.

The fact that the modern notion of sustainable development has been substantially derived from nineteenth century philosophers and political economists such as John Stuart Mill is not, in itself, very surprising. The aspect that defines sustainable development as a uniquely modern phenomenon, is the desire to apply Enlightenment thought to the process of production.

The goal was to arrive at a state of 'stationary development' which would not effect the well-being of future generations, or their ability to reproduce themselves materially. In the nineteenth century, the Enlightenment produced the belief that the expansion of science and rational thought would lead to the development of a society based on 'universal human freedom' (Harvey, 1990: 15). This was particularly the view of August Comte, who sought to apply the principles of rationality to the process of development, and thereby to achieve a stable society based on 'science, technology and industry' (Nisbet, 1973: 239). To Comte, science was a means to 'bring order to societies undergoing radical transformation' (Cowen and Shenton, 1996).

One legacy of the Enlightenment then, was the belief that science and rationality provide the key to human progress, with the result that, since the nineteenth century, science has been called to the service of 'development'. The emergence of state institutions dedicated to the cause of sustainable development was one outcome of this tendency, and a means by which the state encouraged a form of development which attempted to sustain both the ecological and social environments. More recently, however, the notion of sustainable development has taken on a new inclination. No longer primarily concerned with the improvement of society, and in an era dominated by a mode of production which demands constant growth, it is 'development' itself which is now under threat. The environmental devastation and social dislocation which has largely been the outcome of the developmentalist project (McMichael, 1996a; 1996b), has given impetus to calls for a less destructive form of development. Sustainable development has come to mean precisely that, *sustaining development*, or as John Bellamy Foster argues, attempting to maintain the conditions for sustained economic growth (Foster, 1996).

The Brundtland report and sustainable development

While the idea of sustainable development has a longer history than is generally attributed to it, the establishment of the World Commission on Environment and Development and the publication of the Brundtland report (WCED 1987), does stand as a landmark in the recognition of the importance of social justice and environmental issues.

The World Commission on Environment and Development was established by the United Nations in December 1983 with a mission which encompassed four broad objectives:

- to propose long-term environmental strategies for achieving sustainable development by the year 2000 and beyond;
- to recommend ways concern for the environment may be translated into greater cooperation among developing countries and between countries at different stages of economic and social development and lead to the achievement of common and mutually supportive objectives that take account of the interrelationships between people, resources, environment, and development;
- to consider ways and means by which the international community can deal more effectively with environmental concerns; and,
- to help define shared perceptions of long-term environmental issues and the appropriate efforts needed to deal successfully with the problems of protecting and enhancing the environment, a long-term agenda for action during the coming decades, and aspirational goals for the world community (WCED, 1987).

The Brundtland report calls for a 'global agenda for change' and a more sustainable form of development. Defining sustainability as 'development that meets the needs of the present without compromising the ability of future generations to meet their own needs', the report emphasises the necessity for a model of development that addresses the key issue of global poverty. The report stresses the importance of social justice, especially as it applies to the Third World, and gives 'overriding priority' to the need for development which meets 'the essential needs of the world's poor' (WCED, 1987:43). Sustainable development, argues the report, 'requires meeting the basic needs of all and extending to all the opportunity to satisfy their aspirations for a better life' (WCED, 1987: 44). As Goodman (1993: 238) argues, 'the agenda of sustainable development [as defined by the WCED] effectively extends to the poor the human rights of citizenship and participation in society'. To confront the problem of global poverty, the Brundtland report calls for an era of economic growth that is 'forceful', and 'socially and environmentally sustainable' (WCED, 1987).

Despite the Brundtland report's emphasis on the natural environment as a means to 'protect and enhance the resource base', and thus meet 'the essential needs of the world's poor' (WCED, 1987: 43), sustainable development is invariably seen by conservative commentators to be of purely economic interest. For example, according to the economist David Pearce, author of the British Government's *Blueprint for a Green Economy*, sustainable development has come to be interpreted as 'continuously rising, or at least non-declining, consumption per capita, or GNP, or whatever the agreed indicator of development is' (Pearce, cited in Foster, 1996). As Foster (1996: 129) argues, '[s]ustainable development, in these terms, is essentially the same thing as sustained economic growth'. Indeed, in seeking to address global poverty by 'forceful economic growth' the Brundtland report appears to give implicit endorsement to sustained capitalist accumulation.

Sustainability: Application and discourse in the Australian sugar industry

As noted earlier, the model for sustainable development, as articulated in the Brundtland report, is one that is based upon economic growth, balanced by an awareness of environmental issues. Primarily concerned with the alleviation of poverty in developing countries, it is the need for growth in the Third world which is the main focus of the report.

As Goodman has noted however, the notion of sustainability in the context of developed nations, is seen in an entirely different light to the way it is perceived in the less-developed ones (Goodman, 1993: 238). In an era dominated by the doctrine of neo-liberalism and ruled by the need for constant expansion, the principle of sustainable development through 'forceful economic growth' (WCED, 1987), has been embraced with enthusiasm by the more developed industrial nations. Furthermore, as Donati argues, the emergence of an affluent, educated 'new middle class' in developed nations has lead to demands for alternative forms of consumption, and the development of 'niche' markets as the information and leisure-rich members of society seek outlets for political 'participation and self actualisation' (Donati, 1997: 148). One outcome of this tendency has been the developing market for 'eco-tourism'. In response to the demand for places of consumption by eco-tourists, governments in industrial countries have become concerned with questions of 'eco-aestheticism'. Rather than seeking to conserve the environment as a natural resource base for future generations, the provision of, and access to pristine environments as places for consumption by tourism and recreation has become a priority for many governments in the industrial world (Goodman, 1993: 238). In addition, a flourishing market for real estate in non-metropolitan areas has emerged, as disaffected urban dwellers seek to recreate an idyllic rural existence as an alternative to the perceived moral turpitude of urban life. Rather than being worthy of protection for its own sake, or as a resource base for future generations, the natural environment has become an item of consumption by an affluent middle class. As Harvey (1993) argues, viewed in this way, the natural environment has become an exploitable resource and only worthy of protection when it provides some immediate economic benefit to the community.

The Australian sugar industry has not been insulated from the need to conserve the environment as a place of tourist or alternative lifestyle consumption. Indeed, the proximity of many canegrowing areas to culturally significant and environmentally sensitive regions such as the Great Barrier Reef and the Atherton Tablelands, has meant that the industry, to a degree not required of other industries, has needed to be sensitive to the potential danger that sugarcane growing can pose to the environment. The need for environmental sensitivity is, of course, counterbalanced by the view that the viability of the industry, in common with any other which produces commodities within the framework of the capitalist mode of production, is predicated upon growth.

Indeed, the idea of sustainable development has become an intrinsic part of the discourse within the Australian sugar industry, so much so that it has been incorporated into the mission statement of one of the key research bodies, the Cooperative Research Centre for Sustainable Sugar Production. This states that:

> The Sugar CRC will conduct excellent, collaborative and multi-disciplinary research, development and extension to build the knowledge, skills and technology for a sustainable and environmentally responsible Australian sugar industry (CRC Sugar, 1996).

The Bureau of Sugar Experiment Stations (BSES) is yet another organisation devoted to productivity improvement within the Australian sugar industry. A field-day held in 1997, which focused on the theme of 'Sustaining Sugar Production', sought to convey to farmers the principles of sustainable development in the sugar industry, and advised growers that:

> sustaining sugar production with minimal impact on the environment was a major objective of BSES work (Hay, 1997: 2).

The importance that environmental issues holds for the sugar industry was also signalled by the Sugar Environment Forum held in north Queensland in 1998. Notably titled *Sustainable Cane Growing Into the Next Century*, the conference drew together representatives from the major education, industry and research bodies within the sugar industry. According to a senior grower representative, and a significant sugarcane producer in his own right, the forum sought to 'meet the challenge of developing a world competitive export industry which is productive and profitable, sustainable in the longer term and meets community expectations [for] responsible resource management' (Pedersen, 1998).

Addressing the conference, Harry Bonanno, another major grower and Chairman of Canegrowers, the organisation which represents the majority of Australian sugarcane growers, argued that

> [o]ur task is to consider how we can ensure the long-term sustainability of our industry, while, at the same time, achieving the productivity gains which are so essential to our continued efficiency, viability and export competitiveness... Sustainability is more than just another buzz word...[i]t is a guarantee for our future. By continuing to do the right thing by the environment we ensure our on-going productivity and viability, both as individuals and as an industry (Bonanno, 1998).

While prosperous growers appear to give overriding priority to economic success, the literature published by the canegrowers' organisation is a little less transparent. Accordingly, Canegrowers 'recognises that canegrowing must remain, not just economically efficient and viable, but also ecologically sustainable' (Canegrowers, n.d.). Sustainability, for the Australian sugar industry has, therefore, come to involve maintaining a balance between the requirements for economic growth on the one hand, and the need for environmental protection on the other.

However, as Drummond and Marsden (1995) argue, protection of the environment can only take place within the parameters set by the need for profit within the capitalist mode of production. According to these researchers, the idea of sustainability as it has been interpreted by the state and promoted by state institutions, has been 'conceived and promoted within the reflexive progression of capitalism and the conflict and struggles which sustain and renew the dynamism of capital accumulation' (Drummond and Marsden, 1995). Under capitalism, 'forceful economic growth' demands an increase in the productivity of the inputs to the production process. In the case of industrial production, increased productivity has been achieved by the application of technology to the production process, and has invariably come at the cost of decreased need for human labour. For agriculture, the need to increase productivity on the land has, in addition to the diminished labour force, resulted in the increased use of chemicals, and the relinquishment of marginal land to agricultural production.

The process of capitalist development involves the necessity of the progressive development of productive forces. While this process invariably leads to the generation of surplus labour, the increasing impoverishment of the working class and degradation of the natural environment, it does demonstrate the creativity and capacity for innovation, which is latent in humanity. When Marx wrote that '[t]he bourgeoisie cannot exist without constantly revolutionising the instruments of production, and thereby the relations of production, and with them, the whole relations of society' (Marx and Engels, 1967: 83), he spoke not only of the fervour of the capitalist class which propels it to accumulate, but also the creative potential of humankind. Under capitalism, however, the creative capacity of the many is appropriated by the few in the name of productivity, and for the purpose of profit. What is required, therefore, is a model of development which seeks to capture and sustain the creative potential that is latent in all people. As O'Connor (1993) notes, the question still remains as to whether this is possible under capitalism.

Conclusion

As it has been articulated by the various bodies associated with the Australian sugar industry, sustainable development would, at first glance, appear to satisfy the requirements laid out by the World Commission on Environment and Development (1987) for 'economic growth that is forceful, and at the same time, socially and environmentally sustainable'. Closer examination, however, reveals that the limited interpretations of the notion of sustainable development, fail to address adequately the issue of social sustainability which is implicit in the Brundtland report's definition of sustainability. The World Commission on Environment and Development sought, by way of 'forceful economic growth', to give over-riding priority to the alleviation of third world poverty. In so doing, it was clearly placing the interests of people over those of the natural environment. It behoves the Australian sugar industry to seek a model of development that not only sustains the production of sugar and the natural environment in which it is produced, but also the regions and communities that depend upon the production of sugar for their livelihood.

References

Bambrough, R. (1976), 'Introduction' in *Plato: The Republic*, J. M. Dent and Sons: London.
Bonanno, H. (1998), 'Sustainable Cane Growing - Is it Possible?' *Sugar Environment Forum: Sustainable Cane Growing Into the Next Century*: Mackay, 24-25 March.
Canegrowers (n.d.), *Caring for the Environment*: Brisbane.
Cowen, M. and Shenton, R. (1996), *Doctrines of Development*, Routledge: London.
Cooperative Research Centre for Sustainable Sugar Production (1996), *Annual Report 1995/96*: Townsville.
Donati, P. (1997), 'Environmentalism, Postmaterialism and Anxiety: The New Politics of Individualism', *Arena*, No. 8, pp. 147-72.
Drummond, I. and Marsden, T. (1995), 'Regulating Sustainable Development', *Global Environmental Change*, Vol.5, No. 1, pp.1-13.
Foster, J. (1996), 'Sustainable Development of What?', *Capitalism Nature Socialism*, Vol. 7, No. 3, pp. 129-132.
Goodman, D. (1993), 'Scaling Sustainable Agriculture: Agendas, Discourse, Livelihood', in Allen, P. (ed.), *Food for the Future: Conditions and Contradictions of Sustainability*, John Wiley and Sons: New York.
Harvey, D. (1990), *The Condition of Postmodernity*, Blackwell: Oxford.
Harvey, D. (1993), 'The Nature of Environment: The Dialectics of Social and Environmental Change', *The Socialist Register*, Merlin Press: London, pp. 1-51.
Hay, C. (1997), 'Sustainability Vital', *Bush Telegraph*, May.
Marx, K. and Engels, F. (1967), *The Communist Manifesto*, Penguin: Harmondsworth.
McMichael, P. (1996a), *Development and Social Change: A Global Perspective*, Pine Forge Press: Thousand Oaks.
McMichael, P. (1996b), 'Globalisation: Myths and Realities', *Rural Sociology*, Vol. 61, No. 1, pp. 25-55.
Nisbet, R. (1973), *The Social Philosophers: Community and Conflict in Western Thought*, Heinemann: London.
Nisbet, R. (1980), *History of the Idea of Progress*, Basic Books: New York.
O'Connor, J. (1993), 'Is Sustainable Capitalism Possible?' in Allen, P. (ed.), *Food for the Future: Conditions and Contradictions of Sustainability*, John Wiley and Sons: New York.
Pedersen, J. (1998), Address to the *Sugar Environment Forum: Sustainable Cane Growing Into the Next Century*: Mackay.
World Commission on Environment and Development (1987), *Our Common Future*, Oxford University Press: New York.

Index